Trade, Technology and the Ironworking Community in the Iron Age of Southern Britain

Robert M. Ehrenreich

BAR British Series 144
1985

B.A.R.

5, Centremead, Osney Mead, Oxford OX2 0ES, England.

BAR 144, 1985 : 'Trade, Technology and the Ironworking Community in the Iron Age of Southern Britain'.

Price £ 13.00 post free throughout the world. Payments made in dollars must be calculated at the current rate of exchange and $3.00 added to cover exchange charges. Cheques should be made payable to B.A.R. and sent to the above address.

ISBN 0 86054 356 0

For details of all B.A.R. publications in print please write to the above address. Information on new titles is sent regularly on request, with no obligation to purchase.

Volumes are distributed from the publisher. All B.A.R. prices are inclusive of postage by surface mail anywhere in the world.

Printed in Great Britain

TABLE OF CONTENTS

LIST OF FIGURES

PREFACE

The research reported here is intended to increase the understanding of the sophistication and development of British Iron Age blacksmithing technology by metallurgically examining a large number of iron artefacts from a range of areas in southern Britain and placing the results into an anthropological and historical framework. The compilation of the Iron Age database required two years of traveling to museums, sampling artefacts, and performing the metallurgical analysis detailed in Chapter 3. This study is a revised version of my D.Phil. thesis, submitted in July 1985.

I would like to express my gratitude to Professor Barry Cunliffe for all of his encouragement and guidance during the formulation and production of the database and theories presented in this work. I would also like to thank Dr. Peter Northover and Mr. Chris Salter for their instruction and assistance in the metallurgical analysis of the iron artefacts. I am also indebted to both Mr. D. Britton and Dr. H. Cleere for their advice and editorial comments.

The database could not have been assembled without the help and cooperation of Dr. A.J.N.W. Prag of the Manchester University Museum, Mr. F.K. Annable and Dr. P.H. Robinson of the Devizes Museum, Mr. P. Drury and Mr. N. Wickenden of the Chelmsford Archaeological Trust, Ms. E. Lewis and Mr. G.T. Denford of the Winchester City Museum, Mr. W.R.G. Moore of the Northampton Central Museum and Art Gallery, Mr. D. Allen of the Hampshire County Museum Service, Mr. J. Pickin of the Warwickshire Museum, Mr. M. Ridley of the Weymouth and Isle of Portland Museums, Professor W. Manning, Ms. C.A, McArthur, Ms. L. Sellwood, and Mrs. S. Pollard. I am also indebted to Professor C.C. Lamberg-Karlovsky, Mrs. V. Swerdlow, and Mrs. M. Banta for the use of the Harvard University Peabody Museum's VAX 11/730 computer and the PHIND database system. I would also like to thank Dr. Lambertus van Zelst, Mrs. Jacqueline Olin, Mrs. Beverly Smith, and Mrs. Francine Lewis for the use of the Smithsonian CAL facilities. Thanks are also due to Ms. M. O'Hea for assisting with the artefact drawings appearing in this thesis, Mr. G. Wait for his practical advice, and Gus and Jenny Carey for their generous support and understanding over the past three years.

Finally, I would like to especially thank my grandparents Morris and Mary Hasnas, parents Henry and Tema, brother Paul, sisters Beth and Ellen, Alexis Brooks, and Germaine Juneau for their constant support and encouragement throughout the production of this work.

CHAPTER 1: THE BACKGROUND OF THE STUDY

1.1 Introduction

This chapter surveys the history of the study of British Iron Age blacksmithing technology from the earliest published work in 1905 to the present. The progress of metallurgical examination is considered in terms of the methodologies used and the results obtained. Comments on the validity of each previous investigation are also included. The questions raised by the fragmentary character of the conclusions of these earlier studies are discussed in the context of an outline of the present research project and its objectives.

1.2 A History of the Analysis of Iron Age Iron Artefacts from Britain

The scientific study of prehistoric iron and ironworking originated at the turn of the century. The initial objectives of archaeological studies were directed at identifying both the source and the manner of iron's discovery. The localities of primary interest were located in the Near East and the Mediterranean, since these contained the oldest examples of iron. Although an integral part of the study of prehistoric ironworking was the diffusion of iron technology from its source, the absence of evidence of early ironworking in Britain resulted in its omission from the first scientific studies. Near Eastern and Mediterranean artefacts were metallurgically examined in a systematic manner long before barely minimal analogous work was performed in Great Britain. Indeed, only 12 British Iron Age iron artefacts were examined before 1970.

The first analysis of a British Iron Age iron artefact was presented by R.A. Smith (1905) in an article attempting to prove that iron currency bars were the British monetary system referred to by Caesar (V, 12, 4). The results of a brief metallurgical examination of two unidentified currency bars, performed by Professor Gowland, were printed as a postscript to the article. The trace element concentrations of the two bars were:

	C	Si	P	Mn	Ni
A.	tr	.09	.69	nd	.23
B.	.08	.02	.35	nd	nd

Metallographically, the bars revealed very different grain structures. Bar A supposedly resembled meteoric rather than terrestrially produced iron. Gowland states that the high nickel content and low slag inclusion concentration of Bar A proved that its origin was extraterrestrial. Bar B's low nickel content and high slag inclusion concentration implied that it was terrestrially produced, the ore probably originating from the Forest of Dean (Smith, 1905, 194). Tylecote (1962, 209), however, refuted both of these theories 57 years later.

The following year, B.H. Brough (1906) presented a lecture at Glasgow on "The Early Use of Iron" employing Gowland's data, and also that of Dr. L. Beck on a Monzenheim iron bloom. Brough's only statement relevant to the present discussion is:

> Bar A when examined microscopically showed no slag patches, and may have been an exceptional case of the use of meteoric iron, although it contains less than the 6 to 10 percent of nickel usual in such metal. Bar B appears to have been produced by the direct production of Forest of Dean ore (Brough, 1906, 249).

J. Myers published the results of a metallurgical examination of a currency bar from Worthy Down (Hants) in 1922. Trace element concentration analysis of the currency bar revealed:

C	Si	S	Mn	Ni	P
.06	.11	.014	tr	tr	.954

Myers (1922, 133T) concludes from these data that the iron was produced by smelting terrestrial ore.

The currency bar was characterized by a grain structure later determined to be common to these specific artefacts: huge grains, varying carbon concentrations, large elongated slag inclusions, and phosphorus shadows. Myers also subjected the prehistoric phosphoritic iron bar to experiments being performed circa 1920 on similar modern alloys (Stead, 1918; Whiteley, 1920; Whitely, 1921). The phosphorus shadows, according to Myers, were produced by the interaction of phosphorus and carbon during carburization. Supposedly, as carbon diffuses into high-phosphorus iron, the phosphorus recedes. As a result, subsequent decarburization of the metal would reveal remnants of this carbon-phosphorus interaction as shadows. To prove this theory, Myers repeatedly carburized and decarburized a currency bar and observed the pattern produced. Myers states that he was able to reproduce the phosphorus shadows, proving that the prehistoric smith must have similarly carburized and then decarburized the currency bar during its initial production (Myers, 1922, 134T). Both the prehistoric carburizing and the decarburizing processes were deemed accidental by Myers.

Gowland's and Myers' data and theories were reiterated by J.N. Friend (1926, 56-57) in yet another synopsis of the discovery and proliferation of ironworking in prehistory. No theories concerning Iron Age blacksmithing technology were offered.

A decade passed before another Iron Age iron artefact from Britain was sampled. J.A. Smythe in 1936 examined two iron blooms from Sussex. One bloom was discovered in the cinder of the Forewood, Crowhurst bloomery, with no associated pottery. Although Straker (1931, 353) believed it to be Roman, Tylecote (1962, 203) stated that the bloom was manufactured in either the Roman Period or the Early Iron Age.

The high slag content of the Forewood bloom would suggest that it was discarded immediately after smelting. The bloom's carbon content varies from non-existent to as high as 0.3 percent; and the presence of a Widmanstatten grain structure in areas indicates that the bloom had been overheated, possibly to as high a temperature as 1200°C (Smythe, 1936-7, 198).

A gang-chain and a chariot tyre were sampled by R.J. Richardson, G. Richardson, and M. Cook in 1945 and 1946 from the Iron Age lake deposit of Llyn Cerrig Bach, Anglesey (Fox, 1945, 64; Fox, 1946, 75). Trace element analysis of the gang-chain revealed:

C	P	Mn	S
.007	1.05	tr	tr

The Richardsons state that the gang-chain's low manganese and low sulphur concentrations would indicate that the ore from which the iron was smelted was carefully selected for its lack of impurities. The phosphorus concentration of the gang-chain was also shown to be equivalent to that of contemporary wrought iron (Fox, 1945, 64). The Richardsons considered the slag content to be high (Fox, 1945, 64). Prehistoric artefacts, however, generally possess great quantities of slag. The gang-chain's uneven carbon concentration has also subsequently been found to be a common feature of prehistoric artefacts. The Richardsons noted that the scarf weld in the chain was opposite to that produced by a typical right-handed blacksmith. This would suggest that the chain was forged either by a left-handed blacksmith, or by two people, one holding the chain, the other wielding the hammer (Fox, 1945, 65).

The Llyn Cerrig Bach chariot tyre examined by Cook (Fox, 1946, 75) exhibited a carbon concentration estimated to vary between 0.74 and 0.96 percent. Cook states that this high, evenly distributed carbon concentration was the result of a production technique similar to that used for piled or double shear steel. Double shear steel is manufactured by first welding numerous sheets of carburized iron together to form a bar. The bar is then repeatedly cut along the central axis, folded over on itself, and welded to form a finished stock. The tyre's slag content is low, however, this is to be expected in such an extensively forged piece. Trace element analysis of the chariot tyre revealed:

Si	Mn	S	P
.05	.05	.006	.030

In 1962, R.F. Tylecote published <u>Metallurgy in Archaeology</u>. The section on ironworking contains the metallurgical results summarized above together with two additional analyses involving a large bloom from Wookey Hole (0.15% phosphorus and 0.79% carbon), and a currency bar from Bourton-on-the-Water, Gloucestershire (0.35% phosphorus and 0.02-0.8% carbon). The currency bar possesses the common Iron Age grain structure of variable carbon content, high-phosphorus concentration, and large slag inclusions. The bloom, however, is a fine high-carbon steel (Tylecote, 1962). Tylecote was uncertain, though, whether the bloom was deliberately carburized or not.

Tylecote presents various theories concerning the sophistication of iron technology in the British Iron Age, based on the metallurgical analysis of these eight samples. First, he states that the bloom's high carbon concentrations were accidentally produced, noting that partial carburization can occur during production in a bowl furnace. Secondly, he writes that 'the majority of iron required and produced was of the low carbon type' (Tylecote, 1962, 204). Thirdly, he concludes that blacksmiths could carburize the metal when desired, citing the Llyn Cerrig Bach tyre as an example of this. Finally, he notes that the presence of phosphorus in iron would prove beneficial to a prehistoric smith, since phosphorus will harden iron much as carbon will. It is a well known metallurgical fact that while steel can be quenched hardened, phosphoritic iron, i.e. high-phosphorus iron, cannot. Since no quenched artefacts were discovered, however, Tylecote concludes that this shortcoming of phosphoritic iron was unimportant. In conclusion he states: 'We can see therefore, that the high-phosphorus irons produce a high-strength metal without an advanced level of technique and therefore would probably be much sought after by the primitive smith for weapons' (1962, 206).

Unfortunately, these views are mutually inconsistent. Why should a naturally harder iron be of great importance when a technique such as carburization for the production of superior iron is also known? This inconsistency, without doubt, results from the limited number and range of sampled artefacts constituting his database. A credible model of the prehistoric British iron industry requires detailed metallurgical examination of far more than eight samples. Each forging technique cannot be properly ranked in importance given so little evidence of its distribution within the prehistoric ironworking community. A useful model can be devised only after a large number of iron artefacts of varying types and uses have been sampled.

Tylecote's section concerning currency bars presents the data compiled by Gowland, Myers, and one additional bar from Bourton-on-the-water, Gloucestershire (Tylecote, 1962, 209). His first observation is that the high-nickel currency bar sampled by Gowland was terrestrially produced, since the nickel content of the currency bar was far too low for meteoric iron. Furthermore, the ore could not have originated from any of the Northamptonshire ore fields, since these contain low-nickel impurities. The currency bar ore, also, did not originate in the low-phosphorus Forest of Dean ore, since the currency bar's high phosphorus concentration would indicate production from a phosphorus rich iron ore (Tylecote, 1962, 209). The source of the currency bar, therefore, was not identified.

The microstructure of the Bourton-on-the-water currency bar was similar to the other artefacts sampled. The carbon content of the bar varied from low to 0.8 percent. The presence of a Widmanstatten structure, again, revealed that the bar had been overheated. A piled structure similar to the bar Myers analyzed was also discovered. Tylecote states that the low average carbon content of the bar indicates that the carburization was accidental.

Tylecote's final conclusion concerning currency bars was 'that here was a fully developed industry for the manufacture of semi-finished swords and other articles which in time of crisis could be distributed to local smiths and worked up into good weapons' (Tylecote, 1962, 211).

Although elements of Tylecote's work have been criticized here, it must be remembered that his seminal contribution helped define the field of archaeometallurgy. Tylecote's classic, Metallurgy in Archaeology, was the first work to reveal the extent of information that could be gleaned from sampling iron artefacts in its attempt to deduce the level of sophistication of Iron Age blacksmithing technology from metallurgical research. Much of present research is based on the contributions of Tylecote.

In 1968, the first semblance of a systematic metallurgical examination of Iron Age iron artefacts was proposed. J.N. McGrath (1968) analyzed four swords from the Llyn Cerrig Bach hoard in an attempt to add to the general understanding of the manufacturing techniques employed in the production of Iron Age swords (Coghlan, 1956-7; Regioni and Garino, 1955). McGrath (1968, 78) was able to divide the Iron Age swords into six categories (Figure 1.1).

Three of the Llyn Cerrig Bach blades are of the Type 4 category. These blades have a piled structure which runs horizontally, meaning the centre sheet of the pile is stretched between both cutting edges. The core of the Type 4 blades appeared to be lower in carbon than the outer layers (McGrath, 1968, 78). The fourth sword is of a Type 6 piled form. The piles run vertically along the length of the blade, with the centre sheet of the pile forming the entire midrib of the sword, and each successive iron sheet comprising a section of the edges. McGrath states that there is no indication that the core is more or less carburized than the exterior in the Type 6 sword.

McGrath's article concludes with a discussion of the superiority of Class B-4 swords. The determination of the most advantageous sword construction was a significant observation; however, such information is irrelevant if it cannot be shown that prehistoric blacksmiths deliberately selected the Class B-4 form for its superior properties. The currency bars were forged in a Class B-Category 1 sword style also. The swords and the currency bars may have been produced in this manner because the technology and the equipment available in the Iron Age was limited. The thinning of iron sheets before manufacturing a currency bar may have been required either to cleanse the iron sufficiently of slag or to ensure the true weld of the iron. The uneven carbon concentration of both the swords and the currency bars could then be the result of the fact that thin sheets of iron are easily carburized and decarburized accidentally. Speculations concerning the method of currency bar production will be presented in Chapter 5.

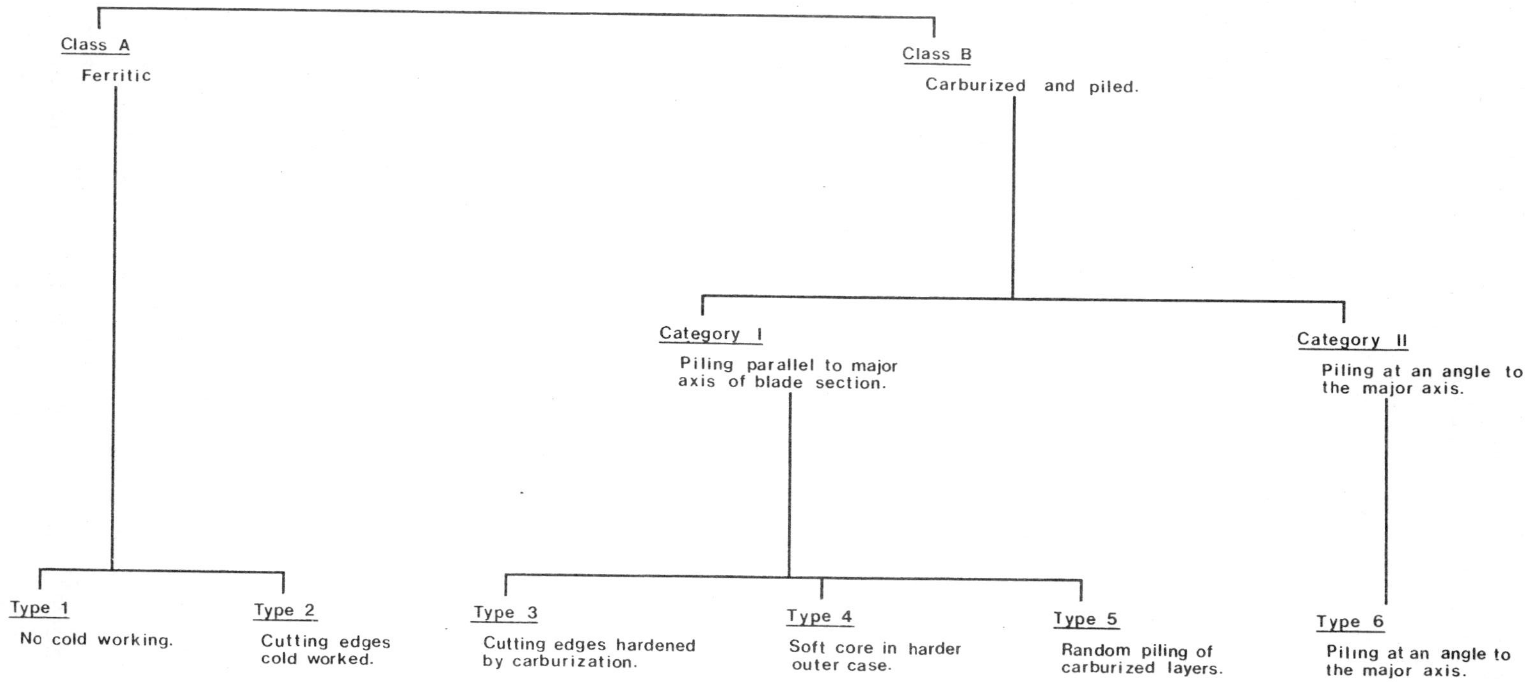

FIGURE 1.1

Classifications of Celtic swords.
[McGrath, 1968, Fig. 4]

In 1970, J.W. Haldane published the first geographically systematic investigation of Iron Age iron artefacts. Haldane attempted to correlate the distribution of iron artefacts in Somerset with their trace element concentrations in the hope of revealing an organization in the production and trade of iron. A total of 29 iron artefacts from seven sites were analyzed. Haldane believed that the results of his study 'showed that there is a high probability that a relationship exists between the trace-element composition and the provenance of ironwork from the Late Pre-Roman Iron Age of Somerset...' (Haldane, 1970, 65). Although this study has been criticized for its assumption that iron is a homogeneous material, Haldane's work more than tripled the number of British Iron Age iron artefacts sampled for trace element concentrations. It is significant because it represents the first comprehensive, and systematic study of prehistoric ironworking involving the selection and sampling of artefacts specifically to test an hypothesis. It is, however, unfortunate that no metallographic examinations were performed.

Haldane's systematic approach had little influence at the time. Studies continued much as they had for the previous 65 years. The next Iron Age artefact sampled was an iron sword from Waltham Abbey (Lang & Williams, 1975, 202-205). Lang and Williams outlined the different technologies available for the hardening of swords using three examples: the Iron Age sword, a Medieval sword from Solingen, and a Frankish sword of the 9th-10th century.

The Waltham Abbey sword possesses a piled structure with the iron sheets varying in carbon content from none to 0.25%. The trace element concentrations determined were:

Mn	Ni	Co	P
.004	.013	.030	.075

Although Lang and Williams selected the Waltham Abbey sword for its variable carbon piled structure, the carbon concentration was neither sufficiently high nor varied for the piling to enhance the properties of the sword. The combination of low-carbon and high-carbon iron strips is desirable for the body of a sword because the properties of the different iron sheets are complementary. A harder metal can be ground to a finer cutting-edge, and is more resistant to wear. One way to increase the hardness of iron is to raise the concentration of carbon. An increased hardness, however, can also cause the metal to be brittle. Low carbon iron has the advantage that it is tough. The metallurgical definition of toughness is 'the capacity of a metal to absorb energy and deform plastically before fracturing' (Davies & Oelmann, 1983, 38). Thus, by combining the two carbon concentrations, a tough blade is produced which can also be honed to a sharp cutting-edge. Lang and Williams, however, do not mention this discrepancy concerning the carbon concentration, and continue to discuss the advantages of a piled blade.

In 1976 C.W. Brewer examined six completely unrelated iron artefacts: an unidentified currency bar from the Dorset County Museum, a Medieval arrowhead, a Saxon spearhead, a Saxon scramasax, a Saxon sword, and a Japanese sword. The currency bar exhibited the usual microstructure described earlier: variations in carbon content from none to low, indications of a Widmanstatten structure, and a piled construction (Brewer, 1976, 1-2). Brewer concludes that since iron could not be melted before the 18th century, different methods were devised for increasing the quantity of carbon in iron. This was supposedly accomplished in the currency bar by piling. No mention is made of the fact, however, that the carbon concentration of the currency bar was far too low to be of use. This study, again, attempted to derive results from minimal, unsuitable evidence.

M.G. Spratling, R.F. Tylecote, P.J. Kay, L. Jones, C.M. Wilson, K. Pettifer, G. Osborne, P.T. Craddock, and L. Biek presented a paper entitled "An Iron Age Bronze Foundry at Gussage All Saints, Dorset: Preliminary Assessment of Technology" at the Archaeometry conference in Edinburgh in 1976. R.F. Tylecote sampled 17 artefacts for this study (Spratling, et al., 1976, 282-291). In this article Tylecote states that the iron sampled was predominantly wrought iron of variable phosphorus concentrations. This would indicate that ore sources of different trace element contents must have been exploited, undoubtedly involving the smelting of local ore and the importation of scrap from distant sources.

Tylecote also writes that deformation twins are evident in one of the seventeen artefacts sampled. These twins indicate that the material had been subjected to heavy striking, a treatment, according to Tylecote, that hardens both high-nitrogen, and high-phosphorus iron more appreciably than modern wrought iron (Spratling, et al., 1976, 254). Tylecote, however, neither identifies the artefact in question, nor states whether such treatment would prove beneficial in this particular instance.

Tylecote then asserts that not only could steel be produced during the Iron Age, but that quenching and tempering were also readily performed. The only example stated of an intentionally carburized artefact was a bronze coated bridle bit. Tylecote also describes bloomery iron as a heterogeneous material which exhibits evidence of some slight unintentional carburization having occurred during smelting and forging. Two artefacts were discovered to be quenched and tempered: a wedge with a piled, variable carbon blade, and an unidentified object.

Tylecote concludes that these 'pieces of iron and steel are very typical of the Iron Age and show, like the chariot tyres of the period, that steel could be made where necessary, and could be quench-hardened to give the higher hardnesses suitable for cutting tools' (Spratling et al., 1976, 284). The chariot tyre Tylecote alludes to is the Llyn Cerrig Bach example only. This conclusion seems unwarranted in view of the fact that only 17 artefacts were sampled from one, exceptional site. Even if Tylecote had compared these data with those obtained from other sites, the results would still be inconclusive. As stated earlier, a meaningful hypothesis concerning the state of iron technology in the British Iron Age requires the sampling of a large number of varying artefact types from various Iron Age settlements. Until this has been done, any theories put forward, however plausible, must be regarded with some scepticism.

This trend of research characterized by limited sampling still continues. An iron bloom sampled by Tylecote from the Little Waltham site is one recent example (Drury, 1978a, 115). Although Tylecote does compare this artefact with other blooms of the same period, the artefact type provides only limited insight into the blacksmithing techniques employed during the Iron Age in Britain.

In 1979, Hedges and Salter published "Source Determination of Iron Currency Bars through Analysis of the Slag Inclusions". They attempted to show that the trace element concentrations of slag inclusions can assist in the determination of the source of iron artefacts. They accomplished this by sampling 25 currency bars from three sites: Danebury, Gretton, and Beckford.

Although Hedges and Salter were mainly interested in the slag inclusions, they also performed metallographic examinations of each artefact. All the bars were found to possess variable carbon contents. Predominantly, the Danebury and Gretton currency bars were low in carbon at the core and had high carbon concentrations near the edges. The Beckford currency bars exhibited a higher carbon content on average; a few of the bars revealed piled structures. The cementite was occasionally spheroidized, indicating an extended heating of the bars at temperatures of between 600 and 700°C.

The phosphorus concentrations of the currency bars were also variable. Predominantly, the Danebury and Gretton bars were high in phosphorus, with mean contents of

0.58% and 0.46% respectively. No phosphorus was detected in the Beckford bars, which would also explain their higher carbon content.

Although this article is not directly relevant to the study of iron technology in the British Iron Age presented here, the methodology is important because it represents a complete and coherent research project based on the sampling of a large number of relevant artefacts.

The "Interim Report on the Ironwork from the Danebury Hillfort" (Salter, 1984), was also mostly concerned with the study of slag analysis. The metallographic results are discussed mainly in relation to the high phosphorus concentrations characteristic of most of the artefacts, and the carbon-phosphorus interaction. These data have been incorporated into Chapter 4.

1.3 The Present Research Project

This publication will attempt to resolve a number of the problems and inconsistencies arising from the fragmentary character of the previous work described above. The methodology introduced here involves the development and analysis of a new database designed specifically to lead to the formulation and resolution of a number of relevant questions.

The first phase in the formulation of the database was the designation of a core area yielding a large number of artefacts for sampling. The core chosen was the hillfort of Danebury (Fig. 1.2). Danebury has produced the most comprehensive, well stratified, and well excavated iron assemblage so far compiled. Three hundred and six iron artefacts were discovered during the first 13 years of excavation. One hundred and forty-one of these artefacts were sufficiently robust to have been metallurgically analyzed. The samples ranged from such tools as awls, hammers, knives, spearheads, and hooked blades, to scrap (i.e. forging debris), broken pieces, and currency bars.

The second phase of the research design involved the sampling of iron assemblages from sites neighbouring Danebury to provide comparative material. All known sites of Iron Age date within a 40 mile radius of Danebury, were identified. The museum or authority housing the artefacts was then approached to determine the possibilities of sampling. In cases where permission was granted, samples were taken from all artefacts determined to be sufficiently robust to withstand sawing.

FIGURE 1.2

The locations of the sites sampled.

It was also of importance to ensure that the iron assemblages sampled in phase two came from small settlement sites as well as hillforts, so that comparison could be made within and between these two levels of the settlement hierarchy. One hundred and thirty-eight iron artefacts were sampled from the hillforts of Barbury Castle, Battlesbury Camp, Buckland Rings, and Bury Hill, and from the settlement sites of All Cannings Cross, Battery Hill Street, Cow Down, Fifield Bavant Down, Little Somborne, Old Down Farm, Rushall Down, West Down-Chilbolten, Winnal Down, and Worthy Down.

The third phase of research involved creating a data-set from other parts of Britain for comparison with the central Southern group. From Northamptonshire, 140 artefacts were sampled from the hillfort of Hunsbury together with five currency bars from Meon Hill. The east of England was represented by 12 artefacts from Little Waltham, Essex. Two of the Continental 'spitzbarren' ingots from the Isle of Portland were also sampled. It was hoped that material from the West Midlands and Somerset could also be included. It was not until the last moment that we learnt that the large assemblage from Beckford could not be made available within the research period. In addition, the Taunton Museum, where the bulk of the Somerset material is housed, found itself unable to cooperate. In the end, only one currency bar from Nadbury was available for sampling.

The fourth phase of the programme was designed to compare the predominantly Early-Middle Iron Age samples with a selected group of Late Iron Age artefacts. Bigbury, Kent, was chosen, providing 24 objects.

In total the database consists of 503 artefacts of differing types originating from different areas of Southern Britain. The artefacts come from a variety of site types and cover the entire period of the Iron Age.

The insufficient quantity of previous metallurgical analyses failed to allow the formulation of convincing theories concerning Iron Age use of carburization, heat-treatment, piling, or even cold-working, since the samples were representative of neither the quantity of Iron Age iron artefacts available, nor of the type of iron artefacts which would reveal advanced metallurgical techniques. With the construction of the present database, the following important questions posed by the previous studies can finally be addressed. First, what was the extent of knowledge of the more advanced techniques for hardening iron, such as carburization and heat-treatment? Were these methods known and employed, or yet to be discovered? Secondly, if the more advanced methods of hardening iron were unknown, what techniques, if any, were Iron Age blacksmiths employing to increase the hardness of iron? Was some other trace element similar to carbon, or was cold-working used to increase hardness? Thirdly, what was the distribution of these techniques within a site's assemblage? Did all the tools requiring superior properties receive special attention, or did only particular tool types warrant property enhancement. Finally, was the distribution of iron metallurgical technology similar throughout Britain, or were there areas, or even pockets, of higher or lower technological capability? The present compilation of an extensive metallurgical Iron Age, database permits an assessment, or at least the beginning of a credible assessment of the level of sophistication and stratification of Iron Age blacksmithing technology.

CHAPTER 2: THE ARTEFACTS AND TECHNIQUES OF SMELTING AND SMITHING

2.1 Introduction

This chapter will summarize the reactions, techniques, and tools involved in the manufacturing of iron and iron artefacts from the ore's collection to the blacksmith's forge. The description of iron production technology will include an over-view of the types and sources of iron ore in Britain. The preparations required to produce a suitable furnace charge, and the smelting procedure will also be discussed. Finally, a reasonably complete account of the evidence of Iron Age iron smelting, as collected by Tylecote (1962, Table 70) and Cleere (1981, 183), will be presented. The summaries of ore collection and iron smelting in the British Iron Age rely predominantly on the work of Cleere (1981) and Salter and Ehrenreich (1984).

The section concerning blacksmithing will introduce the tools and techniques used in the manufacturing of iron artefacts, and present Iron Age examples. The descriptions were compiled from the four volume set *Practical Blacksmithing* edited by M.J. Richardson (1889a; 1889b; 1890; 1891). The existence of numerous, viable, alternative methods for the production of any implement delimits the number of variations of tools and techniques that could be reasonably discussed in this chapter. The following quote from *Practical Blacksmithing* exemplifies the difficulty in standardizing forging techniques:

> Almost every blacksmith has a different style of hammer or handle, and every one thinks that his way of making them is right. One wants a heavy hammer and another a light hammer, for the same kind of work. One wants a long hammer, and another wants a short one. One wants his hammer to stand out and another likes his to stand in...Now, how shall we determine which hammer is the nearest right? I should say both are right, for as long as they can do the work required, and they are satisfied with their hammers, that is all that is necessary (Richardson, 1889a, 199-200).

2.2 Ore Types and Sources

Iron ore is common in Britain. Twenty-nine of the 41 historic English counties have some history of iron smelting. Only nine counties contain no iron ore. The areas now reportedly poor in ore may also have once possessed small quantities, which have long since been mined out. The ore was not, however, always accessible to the Iron Age inhabitants (Salter & Ehrenreich, 1984, 147). Iron ore occurs in geological horizons, some of which were too deep for prehistoric mining to be efficient.

There are four main types of iron ore: magnetite, hematite, the hydrated ores, and the carbonate ores. Magnetite is a ferroso-ferric oxide (Fe_3O_4), and is the richest source of iron. The ore is magnetic, and has a black luster. Magnetite deposits have been discovered in the Devon-Cornwall region, the largest of which is at Haytor, Devon (Collins, 1969, 45-70; Cleere, 1981, 53).

Hematite is a ferric oxide (Fe_2O_3) iron ore, and is generally distributed throughout Britain. Hematite can possess an iron content as high as 60-70% when pure. This ore was also used as a red pigment (Cleere, 1981, 54).

The hydrated ferric oxide ores are classified simply as the hydrated ores. The two most common hydrated ores are limonite ($2Fe_2O_3 \cdot 3H_2O$), and goethite ($Fe_2O_3 \cdot H_2O$). These ores usually occur in bedded masses, such as bog iron (Cleere, 1981, 54).

The carbonate ores constitute the final major ore classification. Examples of carbonate ores are chalybite, and siderite, which contain the metal as $FeCO_3$. The iron content is 40% for the ore within the Weald's Wadhurst Clay formation, and 30% for the Carboniferous clay band (Cleere, 1981, 54).

In addition to the ore types listed above, iron can also form the hard iron sulphide nodules of pyrite or marcasite. These iron sulphide ores usually contain iron concentrations of 45% (Collins, 1969, 82-83).

Although hematite, the hydrated ores, and the carbonate ores are listed individually above, it is rare to find them separately in nature. The formation and history of an ore source affects the types of ore discovered. For example, it is not unusual to find limonite or goethite on the surface of a hematite core, since hematite can be weathered to hydrated ores (Cleere, 1981, 54). Thus, the discovery of two different ore types on a single site does not necessarily indicate that the settlement was collecting ore from two different sources.

The three richest sources of iron ore in Britain are the Weald, the Forest of Dean, and the Jurassic Ridge. The Weald is located on the boarders of Sussex, Surrey, and Kent. The iron ore is discovered as either nodules or thin beds of siderite in the Wadhurst Clay formation, with some oxidized to limonite. Iron is also contained in the Fairlight Clays, the Weald Clays, and the Ashdown Sands; however, the only evidence of prehistoric smelting of these ore sources is at the small settlement site of Runhams Farm, Lenham, Kent (Cleere, 1981, 61; Salter, pers. comm.).

The Weald is undoubtedly the best known of the potential prehistoric ore sources because of the later Roman iron industry based there, but evidence of pre-Roman ironworking is elusive. The sites in, or just north of the Weald which produced small quantities of slag are: Purberry Shot, Surrey (LIA), Hascombe Camp, Surrey (LIA), Brooklands, Surrey (LIA), Ewell, Surrey (LIA-RB), Sedlescombe, Sussex (LIA), Crowhurst Park, Sussex (LIA-RB), and Saxonbury Camp, Sussex (LIA; Tylecote, 1962, 194). Cleere (1981, 183) adds the Late Iron Age, Romano-British sites of Broadfield, Cowpark, Minepit Wood, and Pippingford to the Wealden list. The more sophisticated slag-tapping furnaces dating to the 1st century AD were also discovered at these four sites.

The limited quantity of pre-Roman slag discovered in the Weald would suggest that none of the sites listed above were large iron production centres. This view, however, may be biased by the growth of later iron production sites. Large quantities of slag were hauled away and used by the Romans to metal roads, and metalworkers have periodically stoked blast furnaces with prehistoric slags since the 16th century (Tylecote, 1962, 199; Cleere, 1981, 184-185; Salter & Ehrenreich, 1984, 150-151). Also, since the Roman iron industry was based on the more efficient use of the pre-Roman iron smelting sites, the larger Roman sites would have destroyed all evidence of previous Iron Age iron production, leaving only a restricted and peripheral view of the pre-Roman iron industry.

The Forest of Dean and the Bristol-Mendip region of Britain constitute a second large iron ore source. Limonite is predominantly found in the Carboniferous Limestone, and the Drybrook Sandstone, with some limestone-replacement hematite. The ore occurs in narrow deposits, although occasional larger pockets of ore have been discovered. The concentration of iron averages from 20 to 25% for both the Carboniferous Limestone and the Drybrook Sandstone. In the Bristol-Mendip region, hematite and limonite or goethite have also been found in the Dolomitic Conglomerate. The concentration of this iron, though, is only between 27 and 34%.

The Coal Measures of Southern Wales are located south-west of the Forest of Dean. The Carboniferous Limestone contains large quantities of hematite, and some goethite. Two independent sources of hematite and limonite have been mined within the Taff's Well-Llanharry orefield of Cardiff and Bridgend. The iron concentrations of these two ore sources range from 30 to 55% (Cleere, 1981, 56-57).

The iron smelting remains from the Forest of Dean, the Bristol-Mendip, and the Coal Measures regions of Britain are also limited. For the Forest of Dean, only Bagendon (LIA-RB) on the eastern fringe, and Sudbrook, Mon (LIA-RB) yielded slag. For the Bristol-Mendip region, slag was found at Glastonbury (LIA), Wookey Hole (LIA-RB), and Camerton (LIA-RB); and slag and furnaces were found at Chelm's Combe, Cheddar (c. 250 BC), and Roberrow Warren Cave (LIA). Cleere (1981, 184) states that the quantity of slag found at Camerton would classify it as a large iron production site. Merthyr Mawr Warren (EIA) and Mynydd Bychan are the only two iron production sites discovered in the Coal Measures. Again, absence of remains is undoubtedly associated with the later destruction of iron production sites rather than any actual lack of iron manufacturing in the Iron Age. Cleere (1981, 184) suggests the Forest of Dean region as a possible source of currency bars, although the site of origin still has to be discovered.

The third rich iron ore source is the Jurassic Ridge. The Lower part of the Inferior Oolite, the Upper part of the Middle Lias, and the Middle part of the Lower Lias were all extensively mined into the 20th century. Northamptonshire contains mostly carbonate ores, with portions of the surface weathered to hydrated hematite. Pockets of spathic ores have also been noted. The ores of Northamptonshire are distinguished by their high phosphorus concentrations, yet the iron content is approximately 30%.

Evidence of iron smelting has been discovered at three sites in the region of the Jurassic Ridge: a non-slag tapping furnace neighbouring the settlement of Corby 2 (700-500 BC; Knight, 1984, 165 & 167); a quantity of tapped slag at Castle Yard (c. 450 BC; Knight, 1984, 166 & 188); and a small amount of slag at the Hunsbury hillfort (Fell, 1937, 67 & 95; Tylecote, 1962, 193). The possible discovery of tapped slag at the Early Iron Age site at Castle Yard might indicate that Northants was a more sophisticated iron production centre than the presence of only three sites would suggest. However, again, evidence of iron production was undoubtedly destroyed by later iron-working societies. A recent example of such destruction is the quarrying of Hunsbury in the late 19th century for iron ore. The quarrying destroyed almost all of the centre of the hillfort.

The Iron Age landscape offered additional sources of accessible iron ore distinct from the three areas listed above. The Tertiary Basins of Hampshire and the lower Thames Valley contain mixed beds of clay and iron-bearing sand. Under certain natural conditions, the chemical action of the soil may concentrate the iron into pockets of bog ore or hardpan. The ore source exploited by the inhabitants of Hengistbury Head was produced by such conditions.

The Lower and Upper Greensands also contain iron. Smeltable iron ore has been discovered in the Lower Greensands at Westbury and Seend, Wiltshire. Smelting evidence of this type ore has been found at All Cannings Cross and Swallowcliffe Down. The iron ore characteristic of the Upper Greensands is similar to the Tertiary Basin; the sand is iron rich, and can form bog ore or hardpan under special circumstances.

The Chalklands of Southern Britain contain both pyrite and marcasite nodules. Several kilograms of these iron sulphide ores can still be collected during walks through recently ploughed fields even today (Rod Clough, pers. comm.). These nodules can be purified of sulphur by roasting (Tylecote, 1962, 183; Salter & Ehrenreich, 1984, 147); and desulphurized marcasite nodules can be found in certain regions of the Chalklands (Tylecote, 1962, 199; Cleere, 1981, 53). Both the roasted high-sulphur nodules, and the naturally desulphurized nodules are reducible. These ore sources could account for the smelting evidence discovered at Little Woodbury and Gussage All Saints.

Finally, iron ore is available in Devon and Cornwall. Brixham, Newton Abbot, St. Austell, and Launceston possess small pockets of hematite within the granite quarries. The Great Perran Lode of Newquay is a combination of hematite, limonite, and the carbonates. Four smelting sites were discovered in Devon and Cornwall. Kestor, Devon produced what may be an Early Iron Age furnace; and the sites of Chysauster (LIA-RB), Chun (EIA), and Trevelgue (EIA) in Cornwall all produced slag indicative of smelting. Trevelgue contained a large quantity of slag, some of which was employed in the construction of its Iron Age ramparts (Woolf, pers. comm.). Also, perhaps half a ton of slag was pushed into the ocean during excavation (Woolf, pers. comm.). Trevelgue may well be the only large Early Iron Age iron production site so far discovered.

2.3 Ore Collection

Although the iron ore sources of southern Britain are known, indications of mining remain elusive. The only evidence of Iron Age ore mining is at the site of Lydney in the Forest of Dean (Wheeler & Wheeler, 1932, 32). The absence of mining evidence is associated with two factors. First, mines would be masked by the later collection of ore. As the Iron Age smelting sites were destroyed by subsequent Roman ironworking, so would pre-Roman mines be destroyed by later Roman iron ore collection.

Secondly, the iron ore sources of Britain were so abundant that open-cast collection of ore continued well into the 20th century (Groves, 1952, 5; Cleere, 1981, 132). Pockets of nodular ore and outcrops of deeper bedded ores commonly occurred near the surface of the Weald, the Forest of Dean, and the Jurassic Ridge. This general abundance of readily available iron ore would have made mining unnecessary. Once a pit became too deep for the ore to be easily dug and removed, it would have been abandoned, and a new one begun (Groves, 1952, 5; Tylecote, 1962, 179; Cleere, 1981, 133).

2.4 Ore Preparation

Most iron ore can be smelted immediately on collection. The yield of a furnace, however, can be increased by ore enhancement. The ore should first be examined to ensure that no waste materials are included in the charge. The ore should then be ground or roasted to an optimum size. Large chunks of iron ore will not be thoroughly reduced; and fine iron ore will restrict the air flow within the furnace, decreasing the temperature achieved, and thus preventing the slag from becoming free-flowing, and the smelting from going to completion. Experiments with reconstructions of Iron Age furnaces have determined that roughly 20mm cube chunks of ore produce the optimum yield (Tylecote, 1962, 190; Cleere, 1981, 144).

The ore is reduced in size by crushing or roasting. If the quantity of ore to be processed is limited, the properly sized fragments can be selected by hand. If large amounts of iron ore are required, however, screening can separate it more quickly and efficiently. A wooden screen discovered at an iron production site in Austria was constructed for this purpose (Cleere, 1981, 145). No such implement has been discovered in Britain from either the Iron Age or the Roman period, but this is most likely due to the poor preservation of wood in the archaeological record.

Ore roasting has additional benefits. The hydrated ores, the carbonate ores, and the sulphide ores will form hematite when roasted. Hematite is more easily reduced than the other major ore types. Heating will eliminate excess water in the hydrated ores. For the carbonates, heating will first drive off the carbon dioxide ($FeCO_3 \rightarrow$

$FeO + CO_2$), and then oxidize the FeO to ferric oxide ($4FeO + O_2 \rightarrow 2Fe_2O_3$). If these reactions are not performed before the ore is reduced, they will occur during the smelting operation, robbing time, heat, and efficiency from the actual production of iron. The higher temperature and reducing atmosphere produced within the furnace for the smelting of iron need not be wasted on altering the ore to a reducible state. Thus, the efficiency of the furnace is increased by converting the ores to hematite before smelting.

The sulphur in pyrite and marcasite will not prevent the ore from being reduced to iron. Sulphur, however, is a detrimental impurity. Concentrations of sulphur exceeding 0.035% cause the metal to be brittle and to crumble when forged (Rollason, 1973, 169). Roasting the iron sulphide nodules in an oxidizing atmosphere will eliminate the sulphur from the ore (Tylecote, 1962, 183; Salter & Ehrenreich, 1984, 147). Once the ore is desulphurized, it can be smelted into useful iron.

Roasting is effective at temperatures between 400 and 500°C. These temperatures are easily achieved in charcoal fires, or by burning green wood with the aid of bellows. Since an excess of oxygen is required for roasting, no superstructure need be employed (Cleere, 1981, 146-147). Agricola (1950, 273-274) states that ore roasting was satisfactorily performed by piling the ore on a thick raised base of wooden boards with only a thin layer of sand or charcoal dust covering it to prevent the pile from scattering. Unfortunately, no certain evidence of roasting dating from the Iron Age has been discovered.

2.5 The Smelting Process

The evolution to an iron-based metallurgical society required the resolution of many problems resulting from the higher melting point of iron relative to bronze. The melting point of bronze is approximately 1000°C, which is well below the temperature range of 1100-1300°C obtainable by Iron Age furnaces. The melting point of iron, however, is 1534°C, which is too high for the then existing furnaces.

Iron can be manufactured by heating ferric oxide in a reducing atmosphere. Carbon monoxide will react with the ore to form iron and carbon dioxide ($Fe_2O_3 + 3CO \rightarrow 2Fe + 3CO_2$). The carbon monoxide is produced by the incomplete combustion of charcoal in the furnace. This reaction will occur at 800°C with pure Fe_2O_3. Iron ore, however, is never pure ferric oxide, even after roasting. Large quantities of impurities, such as silicates, will form slag during smelting. In order for iron to be efficiently produced, the slag must be molten, so it can flow away from the reduced iron, and then either settle in the base of the furnace, or be tapped off, depending on the type of furnace employed. The melting point of slag is approximately 1200°C (Tylecote, 1962, 183; Cleere, 1981, 129; Salter & Ehrenreich, 1984, 146). Thus, the temperature of the furnace must be higher than 1200°C to be functional.

Although the slag is molten during smelting, the separation of the iron and the slag is not as efficient as it would be if the iron could also be melted. Slag will still be trapped within the iron bloom on completion of the smelting operation. The slag can be expelled from the bloom by heating it to a temperature where the slag is again molten, and hammering. This 'heating and beating' process will eliminate large portions of the entrapped slag; some residual slag, however, will always remain. Slag inclusions are clearly visible in the grain structure of all finished iron products. A low concentration of slag inclusions is not harmful to the iron, and can even add directional strength. Too high a concentration of slag, however, will make the iron brittle.

A second problem resulting from the low temperatures characteristic of the Iron Age furnaces is that the addition of refractory elements to the charge is impossible. For instance, iron will react with silica during smelting to form the stable slag-phase fayalite (Fe_2SiO_4). This slag phase melts at approximately 1250°C, and removes a large quantity of iron from the finished product. Iron's reaction with silica can be rectified in blast furnaces by adding calcium to the charge. Since both the fayalite and the iron are molten in a blast furnace, the calcium will be able to thoroughly mix with the charge, and replace the iron ions contained in the fayalite, freeing more iron. In the prehistoric furnaces, however, the temperatures were too low to melt the charge. Thus, calcium could not be sufficiently distributed to react properly with the solid slag-phase fayalite. If iron ore with an iron to silica ratio of less than four to one was smelted, most of the iron would be removed by the reaction with the silica, drastically reducing the quantity of free iron produced. Thus, due to the limitations of the furnaces available during the Iron Age, higher quality iron ore had to be used to produce a suitable quantity of iron (Salter & Ehrenreich, 1984, 146).

The common type of furnace discovered in Iron Age contexts is Cleere's type A bowl furnace (Cleere, 1972, 8-11). A bowl furnace is a hemispherical hollow made in the ground ranging in diameter from 0.3m to 1.5m. The hollow is lined with clay and packed with alternating layers of charcoal and iron ore. Experiments have shown that the yield of the furnace is increased if the ore and charcoal are piled well above ground level, and covered with turf or clay. A hole is left on top to permit the escape of smoke, and entrances on the sides of the furnace are required for the installation of tuyeres (Tylecote, 1962, 185; Cleere, 1972, 11). The bellows pump air into the furnace through the tuyeres.

As the temperature of the furnace increases, a reducing atmosphere is created, and the smelting reaction begins. The slag forms and flows to the bottom of the furnace. If no exit has been provided, the slag will collect there until it finally chokes the fire, terminating the process. A furnace which provides an exit for the slag, without allowing excessive heat loss, would increase the reaction time per furnace. Thus, more iron could be produced with no increase in the quantity of ore or charcoal used. Furnaces with exits for the molten slag are termed tapping furnaces, or Cleere's B.1 type (Cleere, 1972, 11-15). It is believed that the tapping-furnace was not introduced into Britain until the 1st century AD (Cleere, 1981, 184). Evidence of tapped slag at Castle Yard, however, may indicate that this more sophisticated technology was actually introduced into Britain at an earlier date. The evidence is still too meager, however, to permit any definite conclusion.

Once the charcoal has burned itself out or the slag has choked the furnace, the top is removed, offering access to the iron bloom. Since the iron was never molten, the bloom is not a cohesive mass, but a porous, slag-rich lump with large cavities where the slag has run out. As stated earlier, the bloom then has to be repeatedly heated and hammered until a large portion of the slag is removed and the iron is an ingot.

The production of iron has been summarized by Salter and Ehrenreich (1984, Fig. 10.1) in a table reproduced as Figure 2.1. The estimates for both fuel and yield were based on a sideritic ore. The final quantity of iron produced will fluctuate depending on both the type of ore employed, and the iron concentration of the ore source.

The most time consuming and demanding of the procedures in Figure 2.1 is undoubtedly the collection of the 90 kgs of fuel required for the production of 1 kg of iron. Charcoal is required for both the smelting and the smithing processes. In the production of charcoal, approximately seven times more wood is needed than the quantity of charcoal desired (Cleere, 1976, 240). Thus, smelting and smithing would have required 420 kgs of wood. Due to the general availability of high quality iron ore during the Iron Age, it would seem that the collection of 420 kgs of wood, and its preparation, would be more time consuming than the collection of the ore itself.

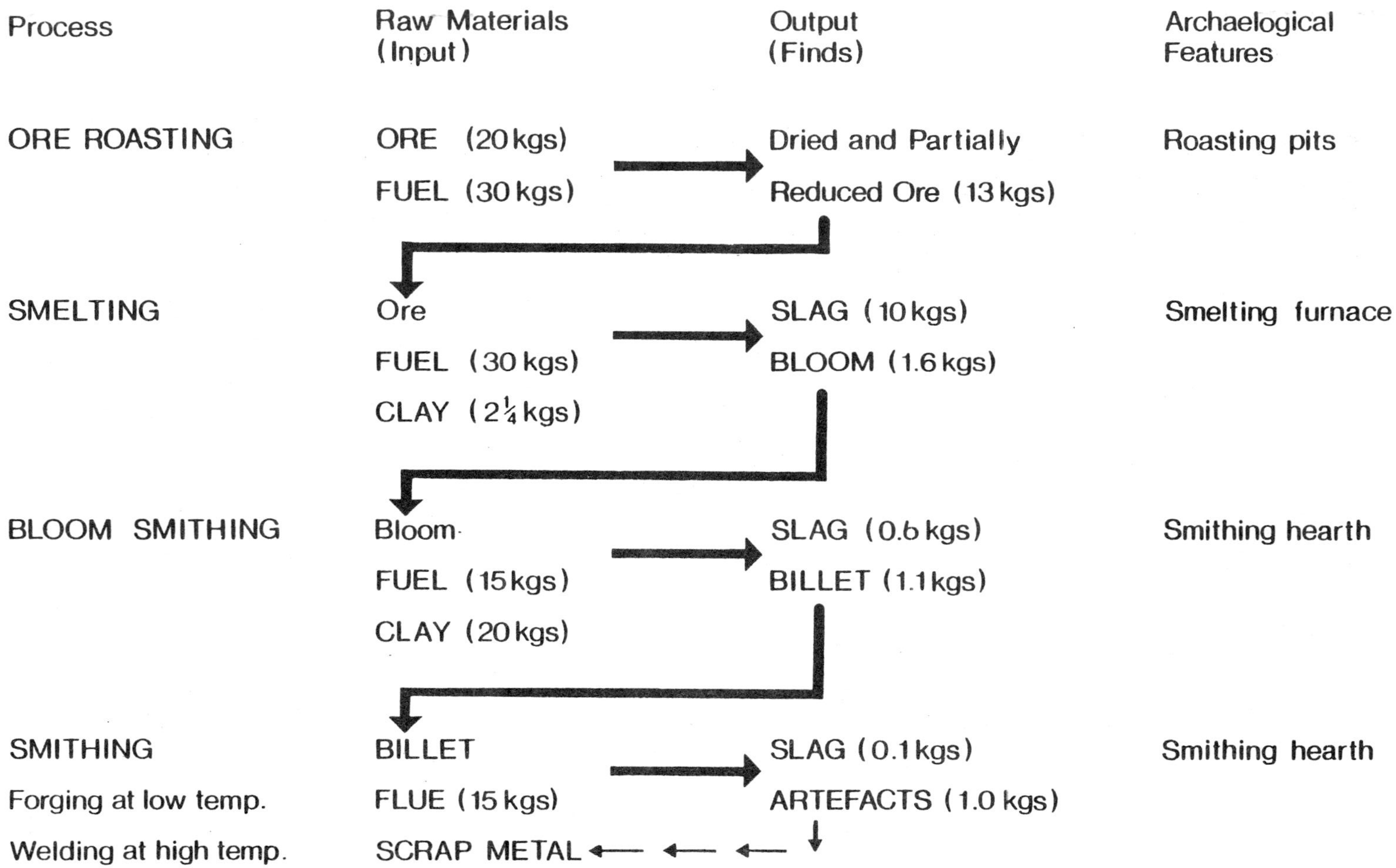

FIGURE 2.1

Summary of the iron production cycle.

[Salter & Ehrenreich, 1984, Fig. 10.1]

2.6 Iron Smithing

The limited temperatures obtainable by the furnaces of the Iron Age not only affected the smelting of iron, but also the manufacturing of iron tools. Artefact production during the Bronze Age was based on casting. Casting allows intricate tools to be made easily. Mass production becomes possible with the advent of permanent molds. Iron, however, could not be melted. All iron artefacts had to be produced by heating the metal until it became malleable, and then hammering it into shape. Thus, the furnace temperature constraint of the Iron Age demanded the development of a new smithing technology for the production of tools and weapons. This section will describe the blacksmithing implements required: hammers, fullers, tongs, chisels, files, and anvils.

The hammer is the most glorified of the blacksmith's tools. The image of the blacksmith deftly forging hot iron into gleaming weapons and tools with hammers has been recounted in endless legends, tales, books, and even films. The requirements for hammers are also the most stringent. The hammer must clearly be harder than the material to be worked. Also, it must be sufficiently tough to withstand extensive striking without cracking or shattering. A technique for manufacturing a fine blacksmith's hammer is to quench the centre of the hammer face to a high hardness, and to decrease the hardness as the edges of the face are approached. This hardness profile makes it less likely for a hammer to crack or split during use. The eye of the hammer should either be quenched to only a slightly lower heat or not be quenched at all. With this treatment, the eye of the hammer can absorb blows without fracturing (Richardson, 1891, 222-224).

As stated earlier, the types of hammers used vary in size, shape, and weight from blacksmith to blacksmith. Six typical hammers are:

500g pene hammer	Used for finishing, light forging, and final chiseling.
800g pene hammer	The common hand hammer.
1100g hammer	A hand hammer used for heavier work and finishing a piece forged with a sledge hammer.
3600g, 5400g, and 7300g sledge hammers	Used for forging heavy pieces. The 5400g hammer is the common form.

Five prehistoric hammers have been discovered: one each from Barbury Castle, Bigbury, and Bredon Hill, and two from Hunsbury. The Barbury Castle, Bredon Hill, and Hunsbury hammers are light, and were probably used for finishing, light forging, and final chiseling. Both the Barbury Castle and the Bredon Hill hammers are now lost. The Bigbury hammer is heavier, and approaches the weight of the common blacksmithing hammer. The head of the hammer is slightly extended, however, which may have made it difficult to control. The hardnesses of the hammers range from 193.2 to 377 VPN, a range sufficient for forging iron without deformation.

Fullers are hammer handled, base struck tools used for elongating iron bars. A fuller set consists of a bottom horn (Fig. 2.2B), and a top, handled tool (Fig. 2.2A). An iron bar is placed between these implements, and the top fuller is struck with a hammer. Forging with fullers will elongate iron without appreciably increasing width. Thus, a long strip of iron can be produced by fullering a thick iron bar of approximately the same width. The horn of the bottom fuller prevents the iron stock from moving during forging. Although the bottom fuller increases the efficiency of the work, it is not absolutely necessary. A bar can still be adequately worked using only the top fuller.

The only extant example of an Iron Age fuller is from Hunsbury (HNY33a; Fig. 2.3). The blade of the implement is 8-9mm wide, and shows no indication of ever having been sharpened. The blade is curved similarly to the bottom fuller, but is socketed like the top fuller. HNY33a is also heavier than present examples, and the butt shows no indication of being struck with a hammer. The tool must have been wielded by an apprentice like a sledge hammer while a blacksmith moved the iron stock in the appropriate directions. Although this tool would work in this mode, the blade would have functioned more efficiently if it had been rounded more. The blade configuration of HNY33a would cause it to penetrate more, spreading the iron laterally to a greater extent than desired.

Other similarly struck, handled tools are the hot and cold chisels. The hot chisel is used for cutting off the ends of hot bars; the cold chisel is used for cutting and splitting unheated bars (Fig. 2.4). Because of the increased hardness of iron at room temperature, the cold chisel requires a harder edge than the hot chisel.

There are only two prehistoric examples of what may be hammer handled chisels: Bigbury's B13 and Hunsbury's HNY20a. Both artefacts exhibit sockets and burrs on the butt of the artefacts from being struck with a hammer. The Bigbury example is robust, and the blade has a hardness of

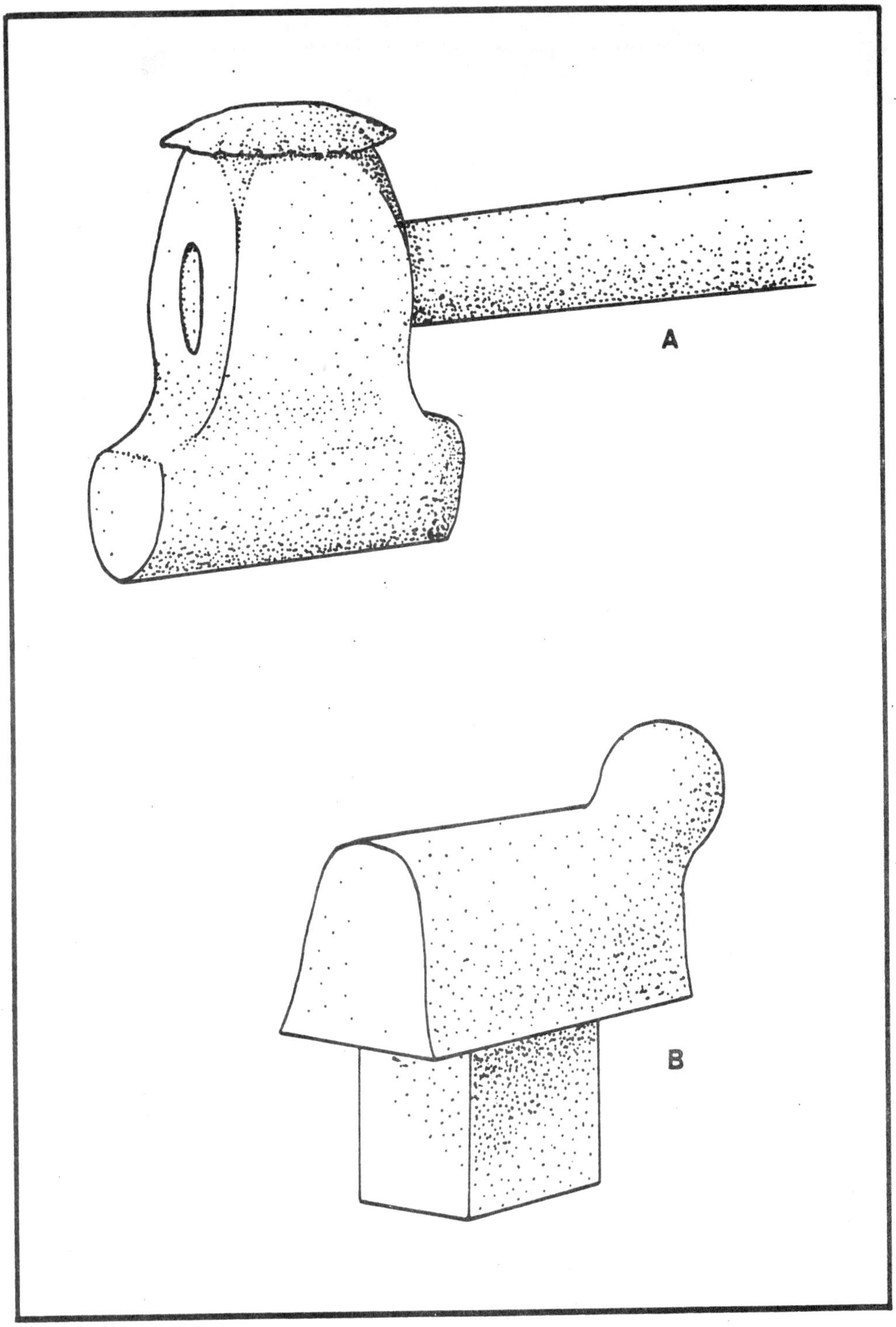

FIGURE 2.2

A typical fuller set
(Richardson, 1889a, Figs. 123-124).

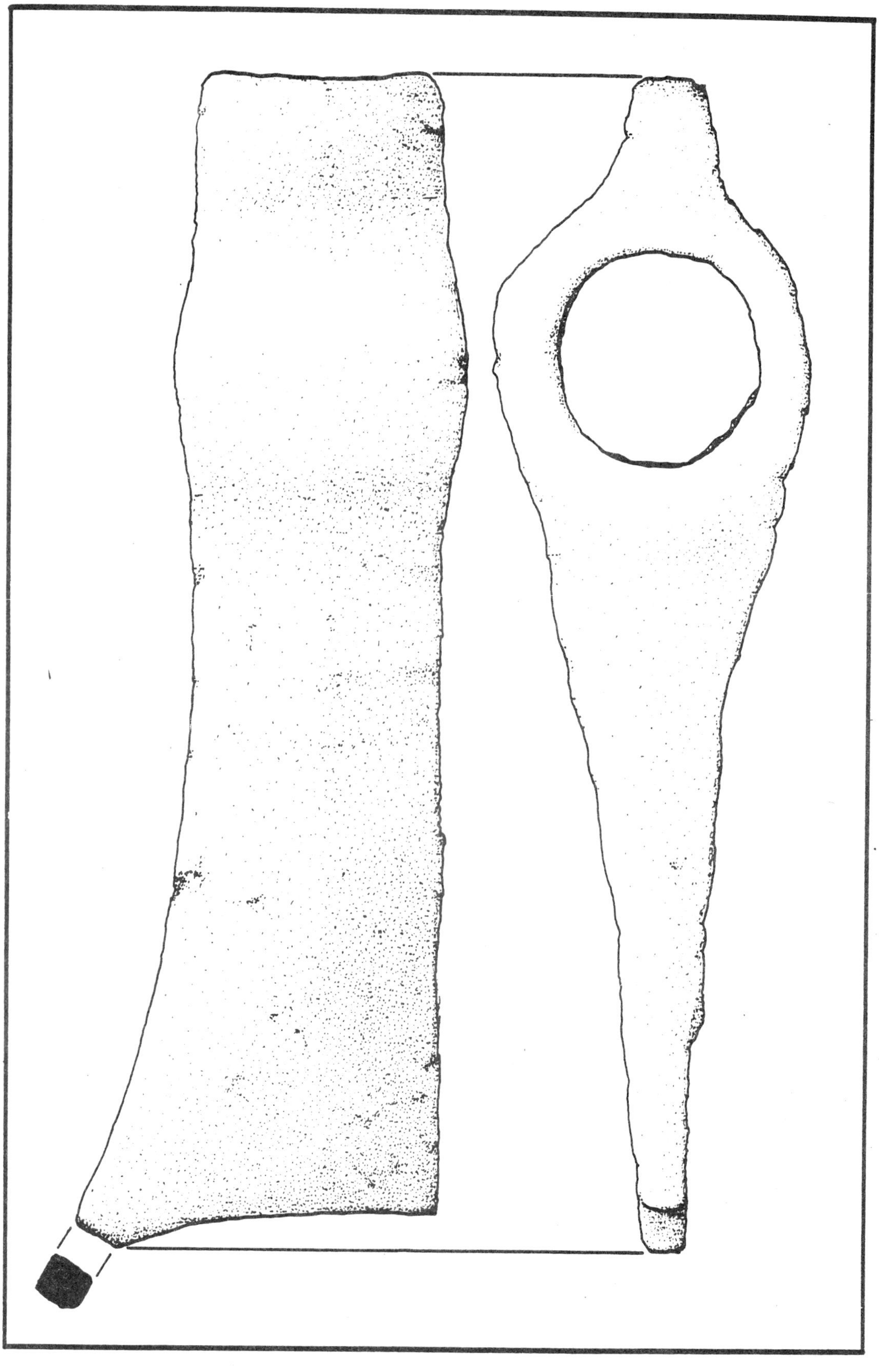

FIGURE 2.3

A possible top fuller from Hunsbury (HNY33a).
Scale 1:1.

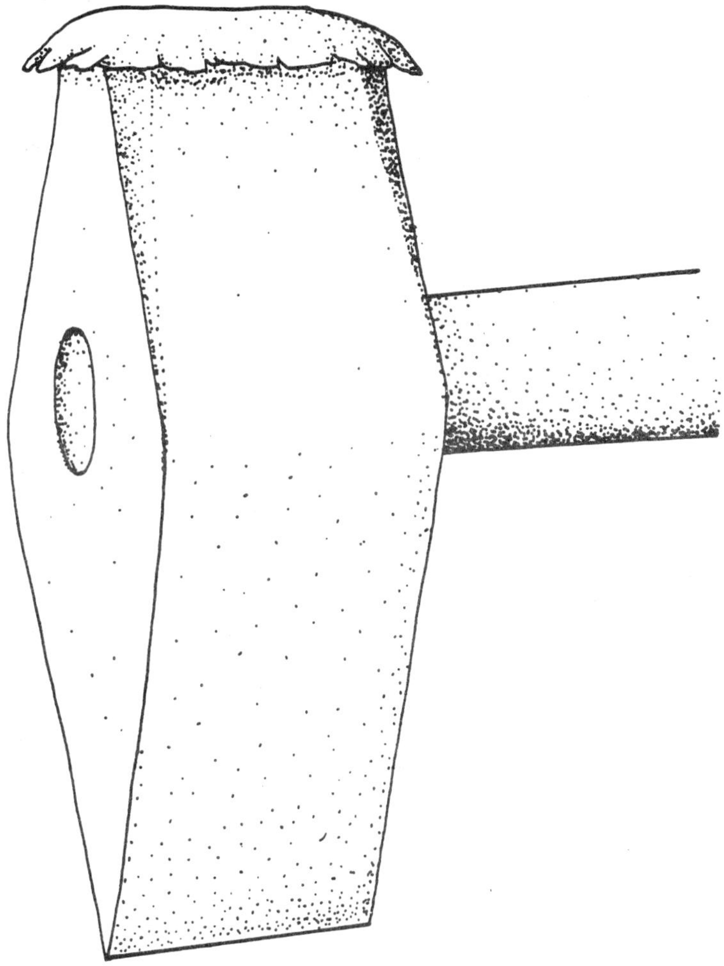

FIGURE 2.4

A typical cold chisel.
(Richardson, 1889a, Fig. 110).

295 VPN. The Hunsbury example is thinner than the Bigbury chisel, but the hardness of the blade is 185.6 VPN, which is still hard enough to cut thin, hot iron.

A number of the wedges could also have been used as chisels. The wedges found at Worthy Down (WD11a; 333 VPN) and Hunsbury (HNY10b; 171 VPN) are both sufficiently strong and hard to cut hot iron. The Worthy Down artefact possessed a quenched and tempered blade, which would have allowed it to cut even cold iron. Thus, certain tools in prehistoric contexts may appear in different shapes than present-day examples. Definite identification in terms of the possible uses of such tools is best obtained from an examination of their microstructures.

The chisels described above are used solely for coarse cutting and not for fine detailed work. Figure 2.5 illustrates two metalworking chisels designed to be used to carve and shape cold iron much as if might be used in the wood craftsman's art. These types of chisels can also be used to split iron when finer control is required. The blade of a chipping chisel should be quenched to a hardness sufficiently high to cut the iron to be worked (i.e. wrought iron), but not too hard that it fractures easily. The back of the blade and the butt should not be quenched, in order to increase resiliency and toughness. A chisel fabricated in this manner will retain a sharp edge, but will also be tough, the hard blade being cushioned by a base of softer, more resilient iron (Richardson, 1890, 149).

Although numerous chisels are found in the archaeological record, the only presently known manner in which a metalworking chisel can be distinguished from a woodworking chisel is by measuring the hardness of the blade. Of the 18 chisels sampled, only six could possibly be used for chipping iron: Danebury's D11a (161.5 VPN), D11b (173.7 VPN), D139 (329 VPN), D157 (309 VPN), and Hunsbury's HNY23b (215.4 VPN), and HNY59a (181 VPN).

Another of the more elaborate and critical of the blacksmith's tools is the tongs. Tongs permit the blacksmith to hold and manipulate hot iron while forging. Figure 2.6 illustrates some of the common types of tongs. Figure 2.6A represents a pair of pick-up tongs. The spring configuration of the jaws exert sufficient pressure when closed to hold small pieces securely. Figure 2.6B represents a pair of flat tongs. These are designed to hold iron plates during forging. The box tongs of Figure 2.6C are for larger bars; the round-bit tongs of Figure 2.6D are designed to hold round iron bars. The ends of tong handles are curved away from each other to allow a ring to be placed around the handle after grasping the metal. This securely clamps the jaws shut and permits the blacksmith to release the tongs without losing grip on the work.

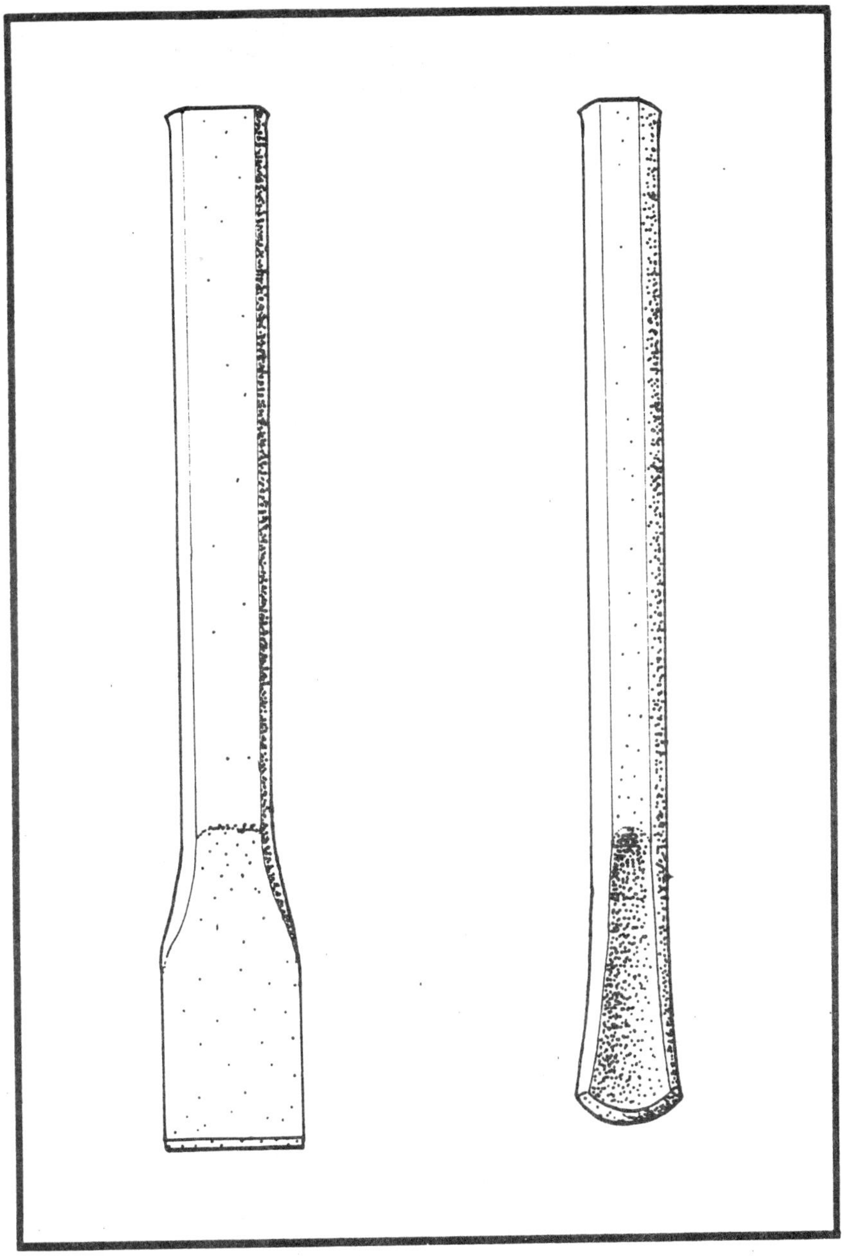

FIGURE 2.5

Two typical metalworking chisels.
(Richardson, 1889b, Figs. 42 & 55).

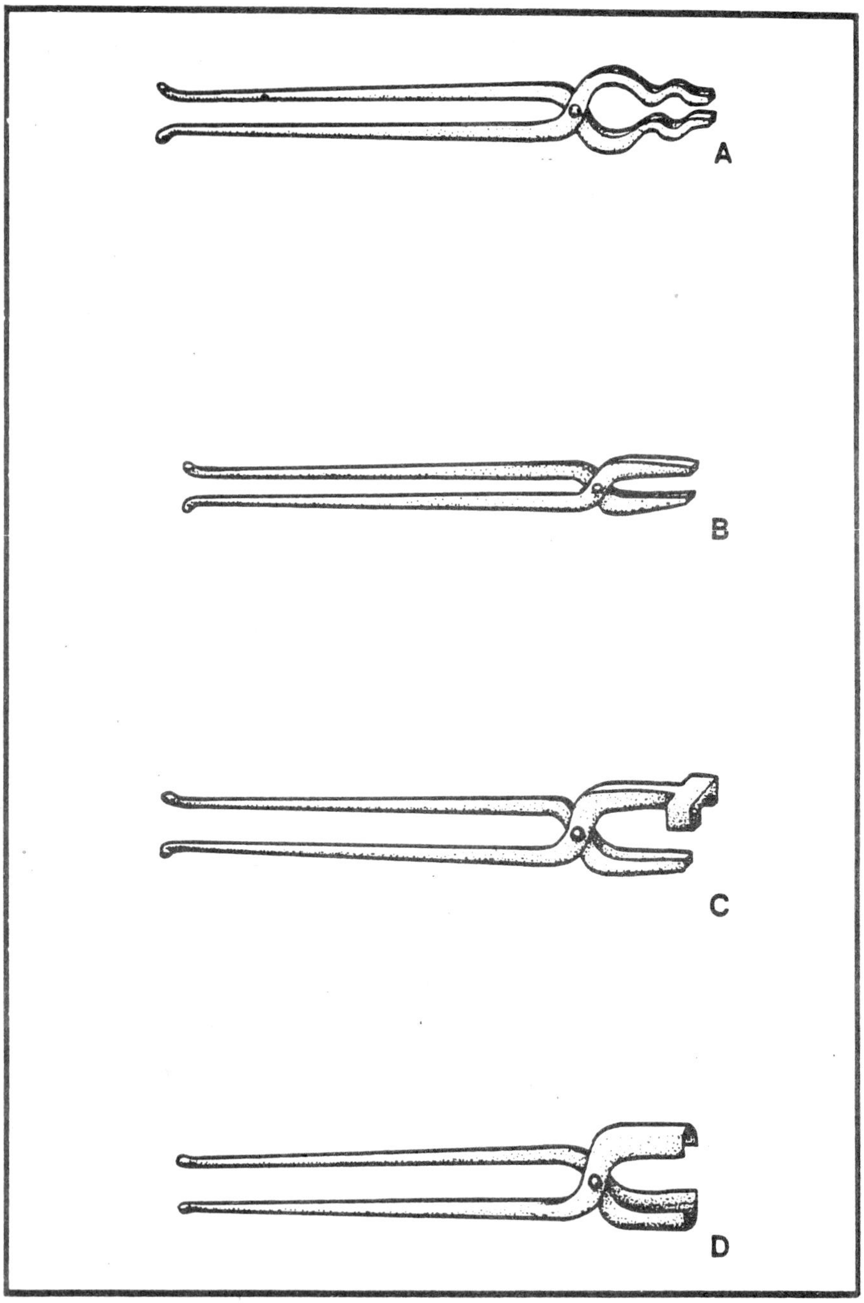

FIGURE 2.6

Four typical tong types.
A. Pick-up tongs C. Box tongs
B. Flat tongs D. Round-bit tongs

(Richardson, 1889a, Figs. 91-93 & 96).

Only three pairs of prehistoric tongs which are similar to present examples are presently extant: two from Llyn Cerrig Bach, and one from Stanton, Norfolk. A fourth pair of tongs of a different configuration were discovered at Bagendon. The Bagendon tongs are similar to large tweezers, and worked by grasping the base. This type of tong, or forceps, resembles the wooden forms employed in India and Africa (Forbes, 1950, 121). Because they are made from green wood, these tongs can handle hot iron without burning significantly. Nevertheless, they do not permit the same grasp and manipulation achievable by using iron tongs. The Bagendon tongs, however, are a reminder that wooden tongs were probably employed in the Iron Age.

The files and rasps used for ironworking are similar in shape and form to those used in woodworking. The only additional necessary requirement of metalworking files is increased hardness. Like chisels, a file needs to be considerably harder than the material being worked, or the file will lose its teeth quickly during use. Thus, files have to be quenched to a high hardness. Of the three files sampled, two are characterized by sufficiently high hardnesses to be used on metal: Danebury's D172 (225 VPN), and Hunsbury's HNY68a (200.8 VPN).

The origins of the anvil are the most difficult of the prehistoric blacksmithing tools to identify. A present-day anvil is made of a hardened steel plate attached to a body of wrought iron. The anvil must be hard, yet tough, and as flat as possible. The heavier the anvil, the less it will move or buck during use.

The only two identifiable prehistoric iron anvils discovered so far are from Bagendon, and Barbury Castle. The anvils resemble large, heavy, iron stakes. Before use, the anvil is driven into a tree stump until it is firmly lodged. Forging can then be performed on the head of the anvil with a minimum of movement. Such anvils are still in use today.

A second possible iron anvil type could be the ingots found at Gussage All Saints (Fig. 2.7) and Hunsbury. These iron objects are flat on both sides, and hooked at one end, forming a loop. They could possibly have been lashed to a tree stump through the eye in such a manner that movement during forging is decreased, while still leaving a suitable surface on which to forge. The large flat area could then be used as a base, and the loop could be used as a horn for fine detail forging and curving iron. The loop on the Hunsbury ingot is split, revealing rough manipulation. Experiments with reconstructions would be useful in providing insight into whether these artefacts are truly anvils.

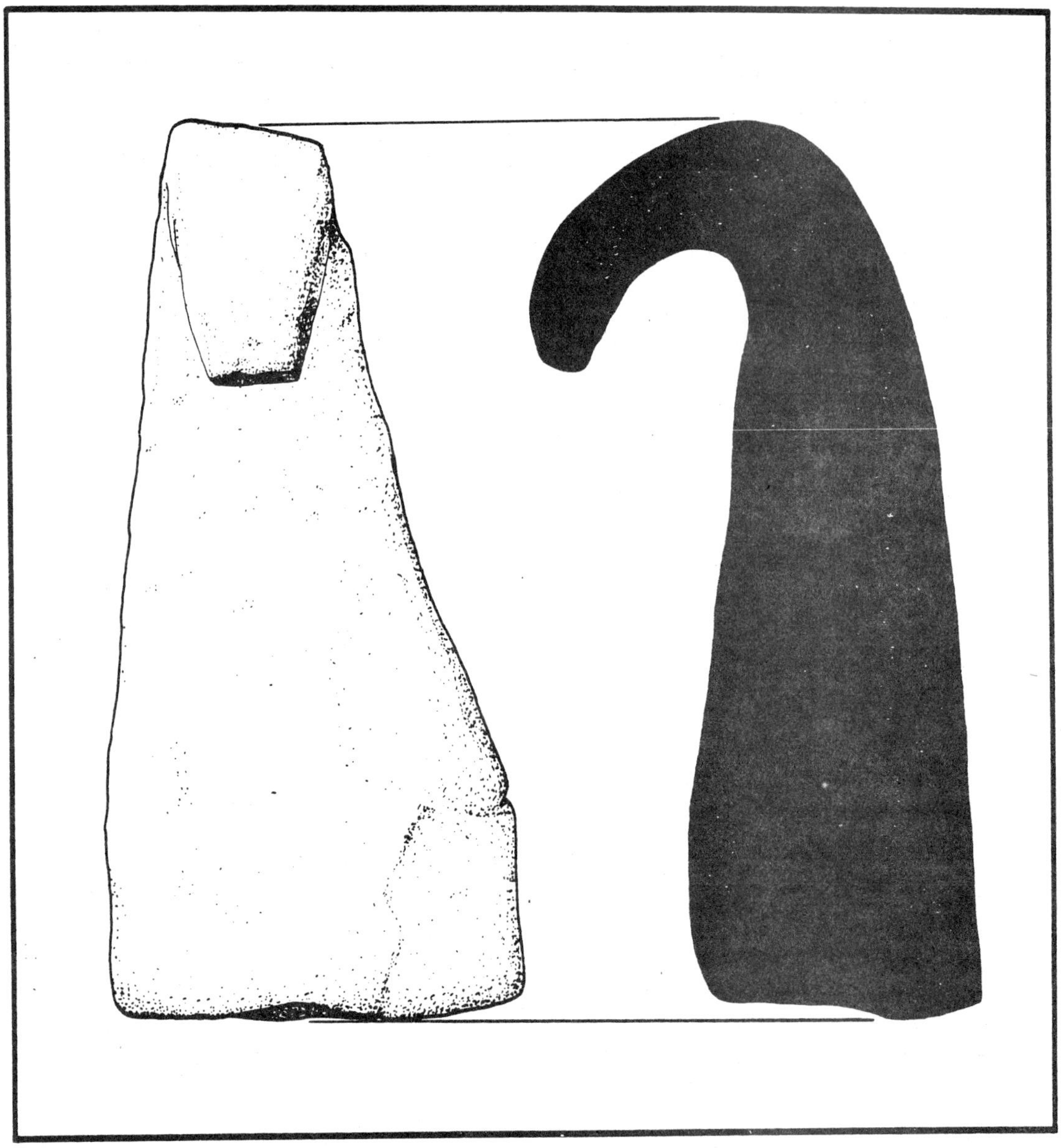

FIGURE 2.7

A possible anvil from Gussage All Saints.
Small Finds 1019.
Scale 1:1.

It should also be noted that evidence of use of stone hammers and anvils exist from Iron Age Britain. A large rounded stone, an approximately $30cm^3$ stone, and a small fragment of a third large stone were discovered with impacted slag particles on their surfaces at Bryn y Castell in Northern Wales. These artefacts constitute a hammer-stone, and two anvils (Crew, 1984, 4). Thus, iron forging tools were not a standard.

The number of blacksmithing tools dating to the Iron Age is limited. Although all of the tools described here could be used for metalworking, their true purposes could as easily have been quite different. The assemblages from Hunsbury, Danebury, Bigbury, and Bryn y Castell are the only ones that can be reliably identified as true black-smithing equipment. The implications of these points will be returned to in Chapter 6.

CHAPTER 3: THE METHODOLOGY OF THE ANALYSIS

3.1 Introduction

This chapter will present the criteria for the selection of the sites and the artefacts examined. The choice of artefacts sampled will be discussed with reference to both the percent of a site's assemblage expected to be sufficiently robust to be examined, and the conditions under which well preserved iron occurs. The procedures for obtaining and preparing the samples prior to metallurgical analysis will also be explained.

A three-part strategy for the collection of metallurgical data was devised to reveal the sophistication of blacksmithing technology and the sources of iron production during the Iron Age. First, the grain structure of each artefact was subjected to microscopic examination to determine inclusion density, grain size, indications of carbon content, and the utilization of heat-treatment and other aspects of production. Secondly, trace element analysis and Vickers Hardness testing was used to exhibit evidence of extenuating criteria which may have affected the production and function of the finished articles. Finally, trace element analysis was employed to identify possible ore sources.

The description of the metallurgical analysis and the data to be obtained is classified according to the specific techniques employed: reflected optical microscopy, trace element analysis using the Camebax Electron Microprobe, and Vickers Hardness testing. Finally, the chronological framework used to order the artefacts will be presented.

3.2 The Selection of Samples

The method of selecting the artefacts to be sampled was summarized in Chapter 1: first, a region of Britain was investigated to determine the range of excavated Iron Age sites; then the authority housing the artefacts was contacted; finally, the iron objects sufficiently robust to be suitable for sampling were selected. The latter consisted, on average, of about half of a typical assemblage. The fluctuations were significant and ranged from 32% for All Cannings Cross to 95% for Hunsbury. The percentage of samplable iron artefacts per site clearly depends on the nature of the soil in which they are embedded. A highly acidic soil and a soil with sulphate reducing bacteria will rapidly corrode iron. Iron can be severely oxidized by electrochemical reactions if buried near an electric generator, or between differing soil types. A well-drained and alkaline soil will produce reasonably preserved iron artefacts. A water-logged and alkaline site will produce exceptionally preserved iron if no salt is present (Rollason, 1973, 145-146).

3.3 The Sampling Procedure

The sampling was performed with a compact drill equipped with a 0.3mm abrasive circular blade. Ordinarily, the saw was used to cut two parallel slits 3 to 4mm apart and 5mm deep into the artefact. A jeweler's saw with a 0.4mm blade was then employed to cut between the slits, freeing a rectangular, parallel-sided sample.

Cutting with the jeweler's saw was found to be time consuming and impractical on objects thicker than 5mm. A more efficient way of sampling thicker artefacts was to angle the two cuts made by the electric saw so as to intersect approximately 6 to 8mm into the artefact. This procedure eliminates the need for the jeweler's saw and produces a wedge-shaped sample.

A parallel-sided sample, however, is preferable for two reasons. First, perpendicular cuts can be continued farther into the artefact, ensuring that a reasonable quantity of intact iron is obtained without having to increase the distance separating the cuts at the surface. Thus, visually visible damage is minimized. Secondly, parallel-sided samples offer the maximum surface area suitable for Vickers Hardness testing. By contrast, the tip of a wedge-shaped sample is easily punctured by a Vickers Hardness tester thereby causing inaccurate results.

3.4 The Sample Preparation

The samples were embedded in labeled, clear plastic resin mounts. Mounting permitted safer storage and facilitated the handling of the samples during preparation and analysis. The samples were ground using progressively finer grit sandpaper to produce a flat, even surface on one side. The first, and roughest sandpaper employed was a 250 grit. The last sandpaper used was an 800 grit. The samples were subsequently polished on a series of rotating cloth wheels impregnated with 6 micron, 3 micron, or 1 micron diamond dust paste. This procedure removes the scratches from the surface of the sample caused by the sandpaper, and leaves a fine mirror polish. Metallurgical analysis followed these preliminary steps.

3.5 The Metallurgical Analysis

3.5.1 Reflected Optical Microscopy - Unetched Samples

The polished samples were examined first under a reflected optical microscope to verify the condition of the surface, and to estimate both the quantity of intact iron and the concentration of slag and forging inclusions.

A well polished iron sample appears bright and featureless when viewed under a reflected optical microscope, since all of the light incident on the sample is reflected by the mirror surface directly back into the eye piece. Scratches and other imperfections produce diffuse reflection, thereby causing dark traces to be visible. Although the fine mirror surface is predominant for all samples upon completion of the polishing process, some surface imperfections are always present. Scratches inevitably occur during polishing as a result of corrosion particles and slag inclusions being dragged across the surface. Samples with too many scratches and imperfections were repolished; a small number of scratches, however, are unavoidable for samples of this type.

Corrosion products and inclusions reflect very little light, and appear dark when viewed under a reflected optical microscope (Fig. 3.1). Remnant images of an artefact's original grain structure can occasionally be observed in completely corroded samples. The wealth of information which eventually may be extracted from these "fossilized" grain structures has been documented by Knox (1963). None of the corroded samples examined for this study of British Iron Age blacksmithing technology, however, exhibited sufficiently convincing images of the original grain structure to permit conclusive identification of the carbon content or grain size once possessed.

As stated in Chapter 2, slag is eliminated from an iron bloom through heating and hammering. The quantity of slag that is eliminated from the iron is determined by the amount of forging. The more an object is heated and hammered, the smaller is the quantity of residual slag. Thus, slag inclusion concentration fluctuates from high quantities in quickly forged tools to low quantities in extensively worked iron objects.

The concentration of inclusions can also be increased during forging. Welding requires the use of fluxes, which decrease the temperature at which iron oxides melt, and facilitate the joining of the metal. These fluxes, however, can be trapped in the iron, producing inclusions similar to slag. Slag inclusions and forging inclusions can be differentiated through trace element analysis. This

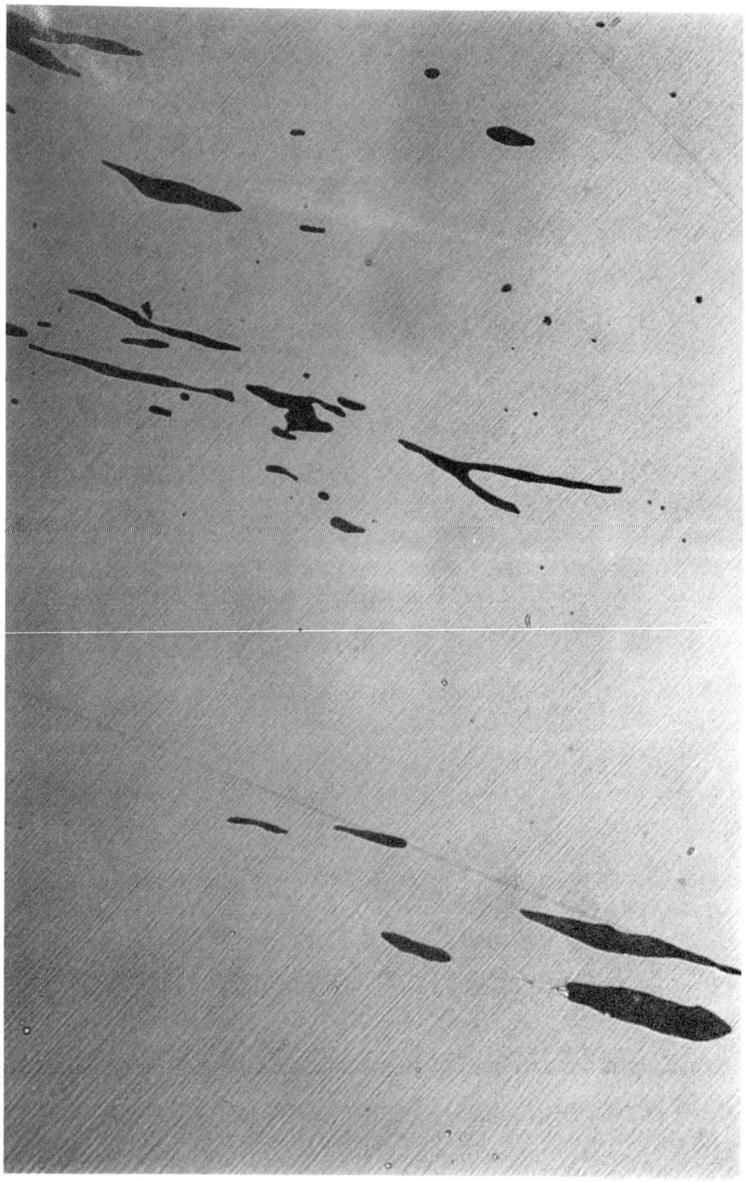

FIGURE 3.1

An unetched sample.
Spearhead D2 - 500X.

process, however, is time consuming, and not directly relevant to this study. Thus, the attempt to distinguish between the origin of the inclusions was not performed.

The presence of inclusions in iron can be detrimental or beneficial depending on the concentration (Rollason, 1973, 117; Davies & Oelmann, 1983, 61; Higgins, 1984, 182). Large quantities of slag and forging inclusions can make iron brittle and of little practical use. A low concentration of directionally oriented slag fibres, however, can impart a directional hardness to the metal, and actually enhance its properties. A correlation between artefact types and inclusion concentrations for prehistoric iron artefacts could be used to show which artefacts received more time and attention during production, and thus, were regarded as more prestigious or important. By contrast the less valuable tools should be identifiable by high concentrations of inclusions, indicating that their production was rapid.

The inclusion density was estimated during examination of the polished samples under a reflected optical microscope. The samples were divided into the categories of low, medium, and high inclusion concentrations. The first 55 iron samples were also photographed, and the inclusion density calculated using lineal analysis. The results of the 55 samples submitted to lineal analysis were correlated with their estimated values to produce the following delineations: "low" contains less than 2% slag and forging inclusions; "medium" an inclusion density between 2% and 4%; and "high" a density over 4%.

The inclusions not eliminated during forging flow as the iron is shaped. The directionality of inclusions in a sample can, therefore, provide insight into the manner of forging. For example, if a square block is reduced in width and thickness, and extended in length, the inclusions will form long, thin fibres stretching along the length of the finished bar. A sample face parallel to the length of the bar would reveal the long, thin fibres. A cross section of the bar, however, would exhibit small, circular inclusion patches, which are the projections of the fibres. Thus, depending on the shape of the inclusions discovered in a sample, the manner in which the iron artefact was forged can be determined.

In actuality, however, it has been found that the determination of inclusion shape is unreliable when studied in two dimensions. An observed thin fibre strand could actually be the cross-section of a wide inclusion plate. Such a broad, thin inclusion could be formed, for example, when an iron cube is reduced in thickness, and increased in both width and length. Also, a small, circular inclusion patch could be the cross-section of a sphere, a possible result of the over-extension and final splitting of a long, thin fibre into numerous small inclusions. One way of differentiating between these possibilities would be to examine two adjoining faces of the sample. This procedure, however, demands a larger sample than would be permissible for this study, since a second face containing intact iron would have to be cut from each sample. Thus, only the slag concentrations were determined for each sample.

3.5.2 Reflected Optical Microscopy - Etched Samples

The determination of the method of production of a prehistoric iron artefact consists of examining an etched sample under a reflected optical microscope. Etching works on the principle that acid will remove more of the loosely held atoms at the iron grain boundaries than those contained in the more strongly bonded iron matrix. After swabbing an iron sample's polished surface briefly with a 0.2% Nital solution, the grain boundaries are scarred, while the grains are only slightly affected. The grain boundaries will then appear as dark lines similar to scratches on examination with a reflected optical microscope. Figure 3.2 shows an etched pure wrought iron sample. The grains could have been produced in Iron Age Britain by either hammering the iron only while its temperature exceeded the recrystallization point, or annealing, which is the extended heating of iron above its recrystallization point. The recrystallization temperature for wrought iron fluctuates depending on the extent of cold-working, the purity of the iron to be annealed, and the duration of heating (Rollason, 1973, 122; Davies & Oelmann, 1983, 58-59; Higgins, 1984, 329-330). Heating an iron bar at a low temperature for an extended period of time will frequently have the same effect as heating to higher temperatures for shorter durations. For example, a steel bar which can be annealed at 675°C in 8 minutes, would require 32 hours at 550°C to achieve the same grain size (Rollason, 1973, 122).

The grain size of an artefact can be controlled by regulating the amount of annealing performed. If iron is heated to a temperature just above the recrystallization point and then cooled quickly, the grains of the object are permitted to nucleate, but not grow. Thus, a very fine grain structure is produced. The grain structure of the artefact exhibited in Figure 3.2 reveals that the object had been well annealed. The grains had not just been allowed to nucleate, but also to grow to a very large size. The grain size of an artefact affects its properties. An object with large grains is softer and more ductile than a fine grained one.

Initially, it was hoped that correlations of grain size and artefact type could be attempted. Microscopic examination of the samples after etching, however, revealed

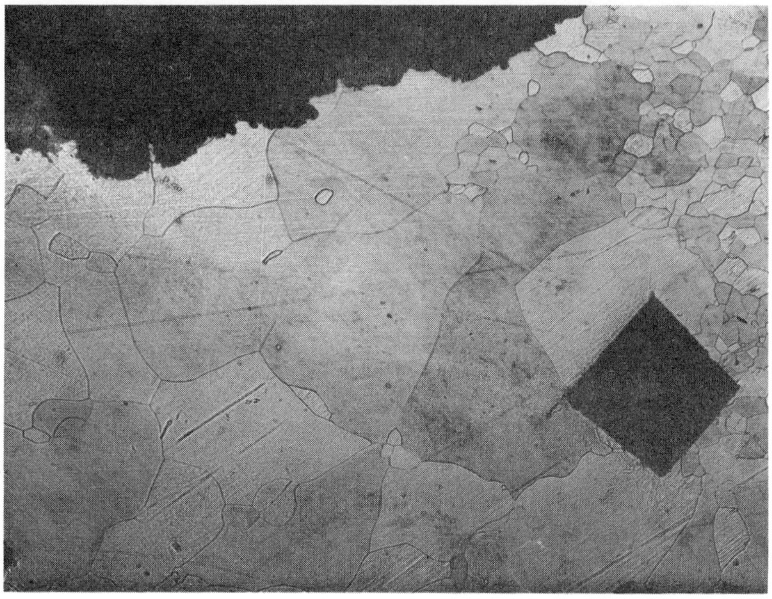

FIGURE 3.2

Wrought iron.
Nail WND5b - 80X.

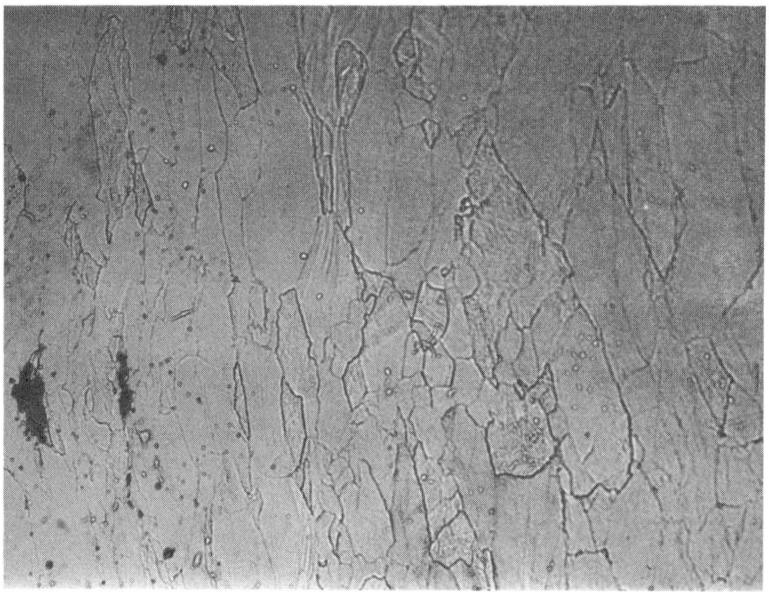

FIGURE 3.3

Cold-worked iron.
Nail WND5a - 350X.

that grain size did not accurately represent the annealing time employed. Phosphoritic iron grains tend to be larger than those of pure iron, even within single samples. Thus, the correlations produced were not of grain size versus artefact type, but grain size versus phosphorus concentration. The establishment of a standard means by which the annealing times of high-phosphorus and low-phosphorus iron could be compared proved impossible. All comparisons of the different metal types' grain structures were inaccurate, and thus, had to be ignored.

If an iron artefact is hammered while at a temperature below the recrystallization point, the grains are deformed and elongated in the direction of the forging (Fig. 3.3; Davies & Oelmann, 1983, 57). This process is known as cold-working. Cold-working alters the lattice arrangement of the grains and prevents the movements of dislocations. The ductility of iron is reduced when dislocations are pinned, and the metal becomes harder. Although the effects of cold-working can be easily rectified by annealing, the increase in hardness, especially in the direction of the grain elongation, can be beneficial. The cold-working of a cutting-edge is a simple method for increasing its hardness and enhancing its ability to be honed to a sharper blade.

In this study, cold-working was identified during examination of the etched samples under the reflected optical microscope. The presence of either grain elongation or deformation twins in a sample was a definite indication that the artefact was subjected to cold-working. Cold-working was studied as a possible method for deliberately hardening iron to enhance the properties of certain artefacts. Thus, the condition of an object's grain structure was ascertained first on examination of an etched sample under a reflected optical microscope.

The addition of carbon to iron forms steel, a harder iron alloy. At the low carbon concentrations attainable during prehistory, the higher the concentration of carbon in iron, the harder the metal (Fig. 3.4). An increase in carbon content, however, also increases the tensile strength of iron, and decreases the amount the metal can be elongated before fracturing. Tensile strength and percent elongation are both indicators of ductility. An increase in tensile strength with a decrease in percent elongation indicates a reduction in ductility. Thus, with an increase in the carbon concentration in prehistoric iron, its hardness will be raised, enhancing both its the ability to be honed to a sharp edge and its resistance to wear, but the metal will also be less resilient and more brittle than wrought iron.

High-carbon iron appears as dark patches under a reflected optical microscope (Fig. 3.5). These dark areas are actually the products of alternating light and dark

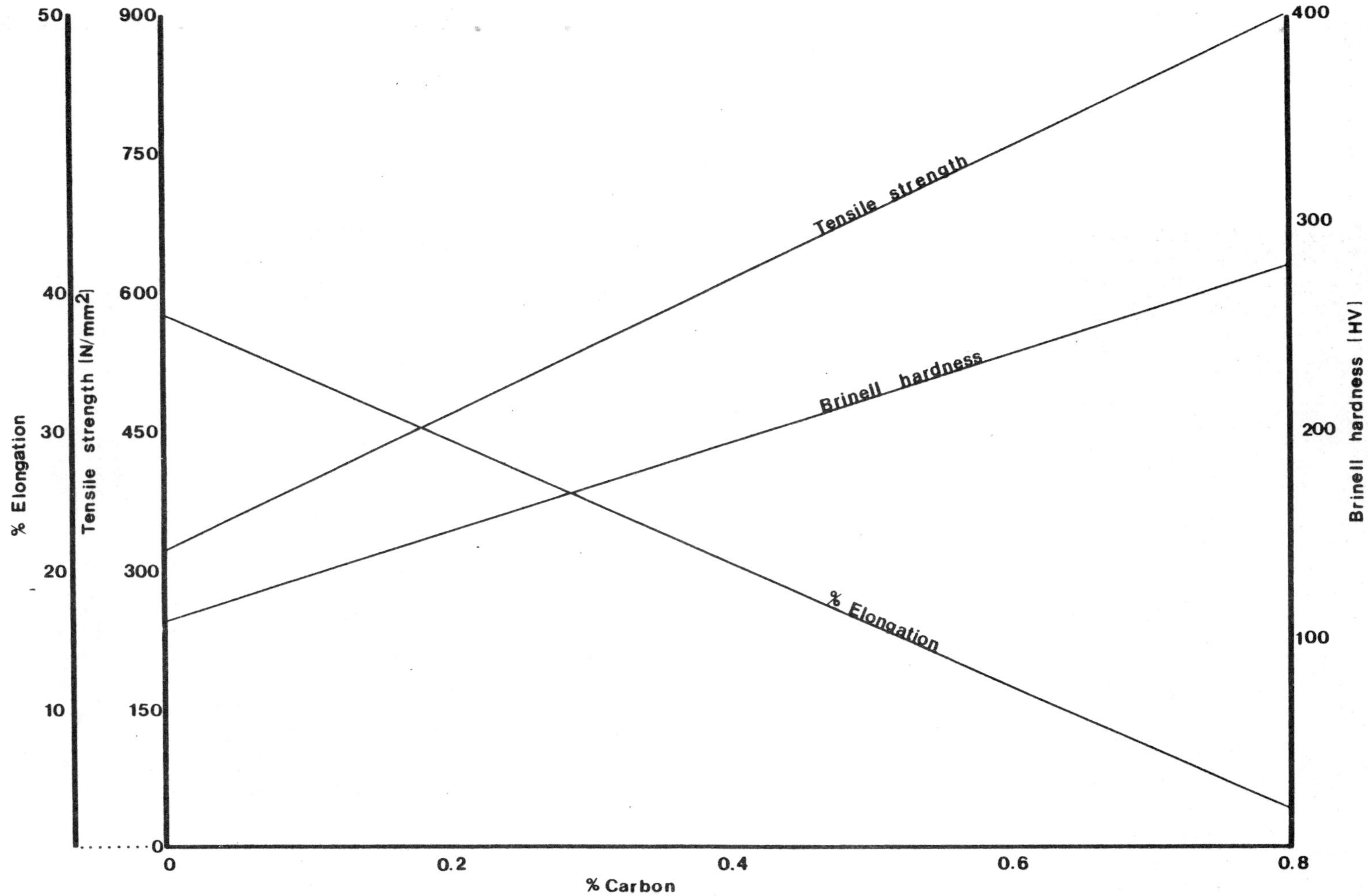

FIGURE 3.4

The mechanical properties of steels as a function of composition.

(Davies & Oelmann, 1983, Fig. 7.5)

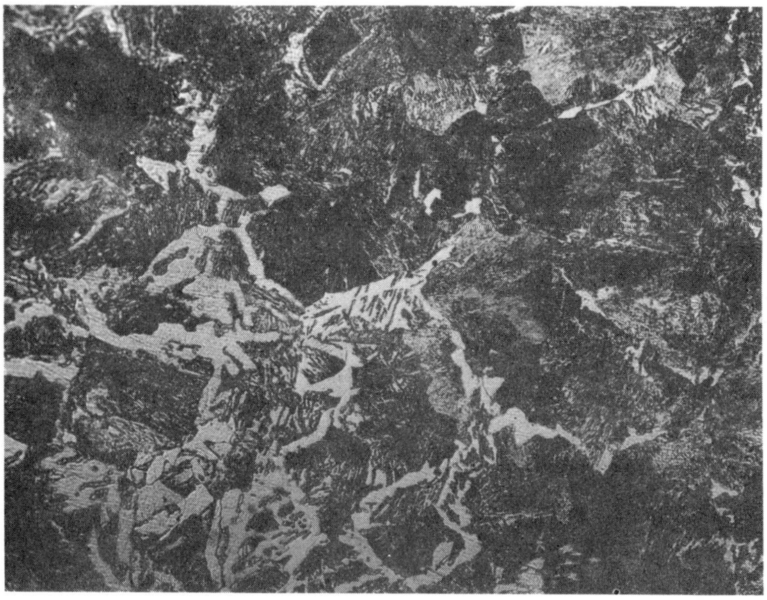

FIGURE 3.5

Pearlitic iron.
Strip D22a - 350X.

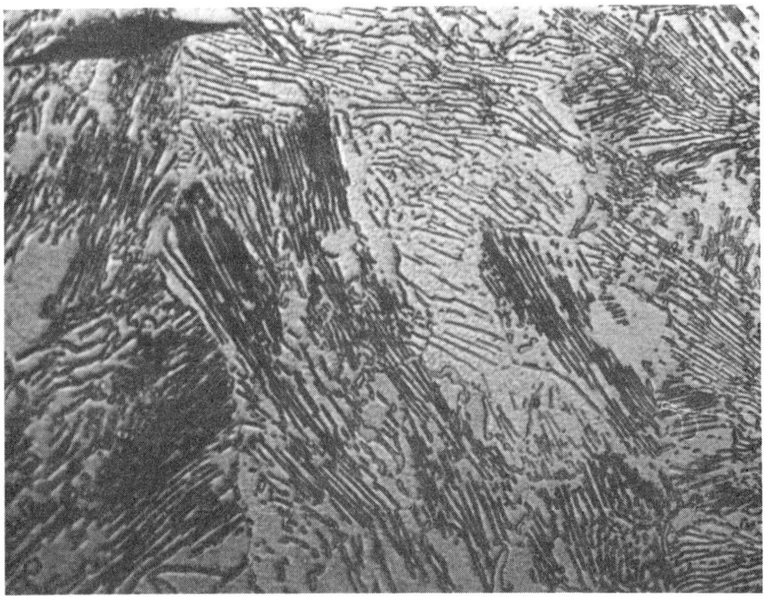

FIGURE 3.6

Pearlitic iron.
Strip D22b - 800X.

lines (Fig. 3.6). The lighter bands originate from the hard, high-carbon iron alloy of cementite, and the darker lines originate from pure ferrite. This common structure is known as pearlite, due to its 'mother-of-pearl' luster.

If steel is heated to a temperature of 680-690°C for long durations (approximately 8 to 24 hours), the carbon distribution in iron will alter. The cementite forms small spheroids in a pure iron matrix (Fig. 3.7). This process is known as spheroidising, and results in greatly softened steel. Spheroidised iron can be rectified by quenching and tempering (Rollason, 1973, 181; Davies & Oelmann, 1983, 109-110).

During the Iron Age, steel could be produced only by carburization. Iron is carburized by packing it in a rich source of carbon (i.e. charcoal, organic materials, etc.) in a furnace with a reducing atmosphere for extended periods at temperatures of 900 to 950°C (Rollason, 1973, 204; Lang & Williams, 1975, 199; Davies & Oelmann, 1983, 114-115). Under these conditions, iron will absorb the unoxidized carbon from the atmosphere of the furnace. The rate of diffusion of carbon in iron during carburization fluctuates according to the type of steel, the temperature of the furnace, and the carburizing mixture. At 900°C, however, iron can be carburized to a depth of approximately 0.2mm in an hour. This process can be performed deliberately by a blacksmith, or accidentally during smelting or forging under appropriate conditions. Piaskowski (1974, 325) speculates that evenly carburized artefacts could also have resulted from the selection of those areas of iron blooms naturally high in carbon by Iron Age blacksmiths, and not due to deliberate carburization. High-carbon iron could have been identified by a blacksmith because its increased hardness would have made forging more difficult.

Poorly carburized iron may be caused either by the incorrect placement of an iron bar in a forge, causing uneven carbon distribution or surface decarburization, or by phosphorus impurity bands. Impurity bands high in phosphorus slow the diffusion of carbon in iron and cause layers of high-carbon, low-carbon, and high-phosphorus iron to be formed (Fig. 3.8; Boylston, 1928, 446; Rollason, 1973, 170; Davies & Oelmann, 1983, 86).

The distribution of carbon in an artefact can be either beneficial or detrimental depending upon the artefact's purpose. For example, a blade with a steel edge is superior because the additional carbon content would increase its hardness, allowing it to be honed to a finer, wear-resistant cutting edge. A chain, however, would be adversely affected by a high carbon content, since the increased hardness would decrease its ductility and toughness. Thus, the carbon concentration of an iron tool can either help or hinder its application.

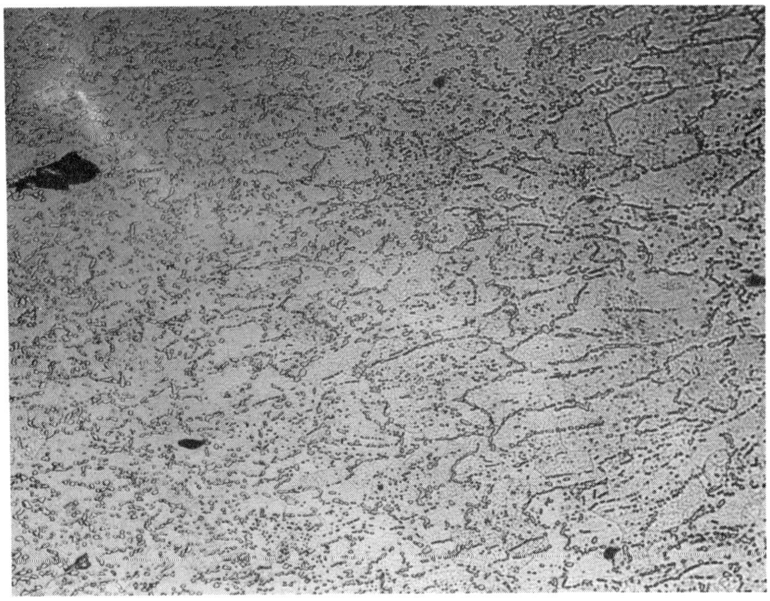

FIGURE 3.7

Spheroidized iron.
Rod HNY24b - 350X.

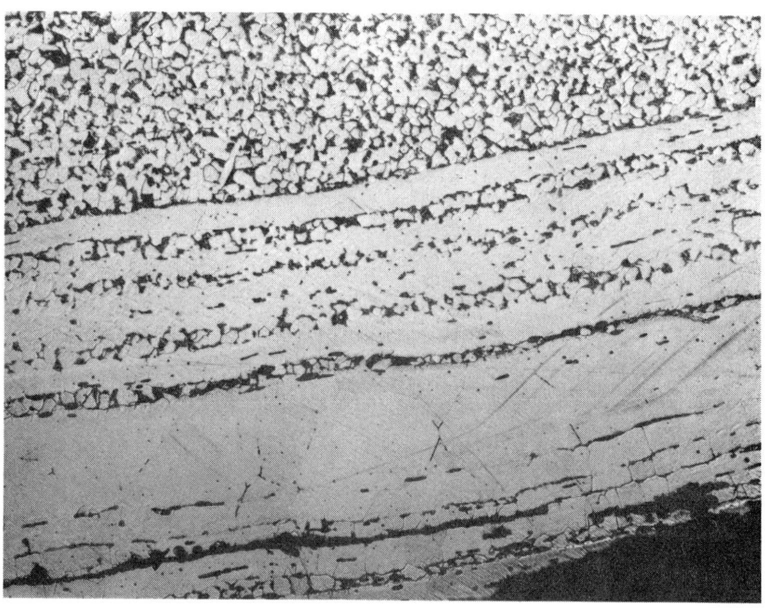

FIGURE 3.8

Banded, phosphoritic iron.
Spearhead BC4b - 50X.

The quantity of carbon in iron is estimated during microscopic examination of an etched sample. Examination of the distribution of carbon in an artefact can show whether Iron Age blacksmiths understood either carburization or bloom selection, and, if so, how they applied this knowledge. Piaskowski (1974, 324) states that it may be concluded that Iron Age blacksmiths were aware of naturally carburized materials or carburization if any of the following conditions apply: (1) tools are regularly discovered with low-carbon and high-carbon iron welded together; (2) a certain tool or weapon is consistently found possessing a high carbon concentration; (3) the combination of low-carbon and high-carbon iron is discovered in the superior blade configuration discussed in Chapter 1 (Fig. 3.9); or (4) a tool contains a higher carbon concentration in only those areas requiring an increased hardness.

Steel with a carbon content in excess of 0.3% can be significantly hardened by quenching. Quenching is the rapid cooling of steel from a temperature of approximately 750°C to approximately 50°C by submerging in a liquid. The grain structure produced by quenching is determined by the speed of temperature reduction. A steel bar which is cooled from 750°C to 50°C in approximately one second will display a pure martensitic structure. Martensite is easily recognized under a reflected optical microscope due to its angular, needle-like appearance. The hardness of martensite can range from 300 VPN for a 0.3% carbon quenched steel to 900 VPN for a 0.7% carbon quenched steel.

Two other quenched structures are upper bainite and lower bainite. A pure upper bainite grain structure is formed by quenching a steel bar from approximately 750°C to 450°C in under one second, and then allowing the bar to cool slowly. Lower bainite is produced by first quenching a steel bar from 750°C to approximately 300°C in under one second, before allowing the bar to cool slowly. The hardness of upper bainite can range from 550 to 650 VPN. Lower bainite ranges in hardness from 450 to 550 VPN. Although the bainite grain structures are difficult to determine accurately under a reflected optical microscope, the correlation of the artefact's carbon content and hardness can reveal if an object has been heat-treated. Figure 3.10 is a graph of carbon concentration versus hardness with delineations for the different quenched structures.

Quenched structures are very hard, but also highly stressed and very brittle. Tempering is usually required before the metal can be used, or else it is liable to crack and distort. Heating quenched steel to a temperature of between 450 and 650°C for an extended period of time will relieve the internal stresses, and reduce the material's hardness, but, more importantly, will increase its toughness (Rollason, 1973, 186; Davies & Oelmann, 1983, 112).

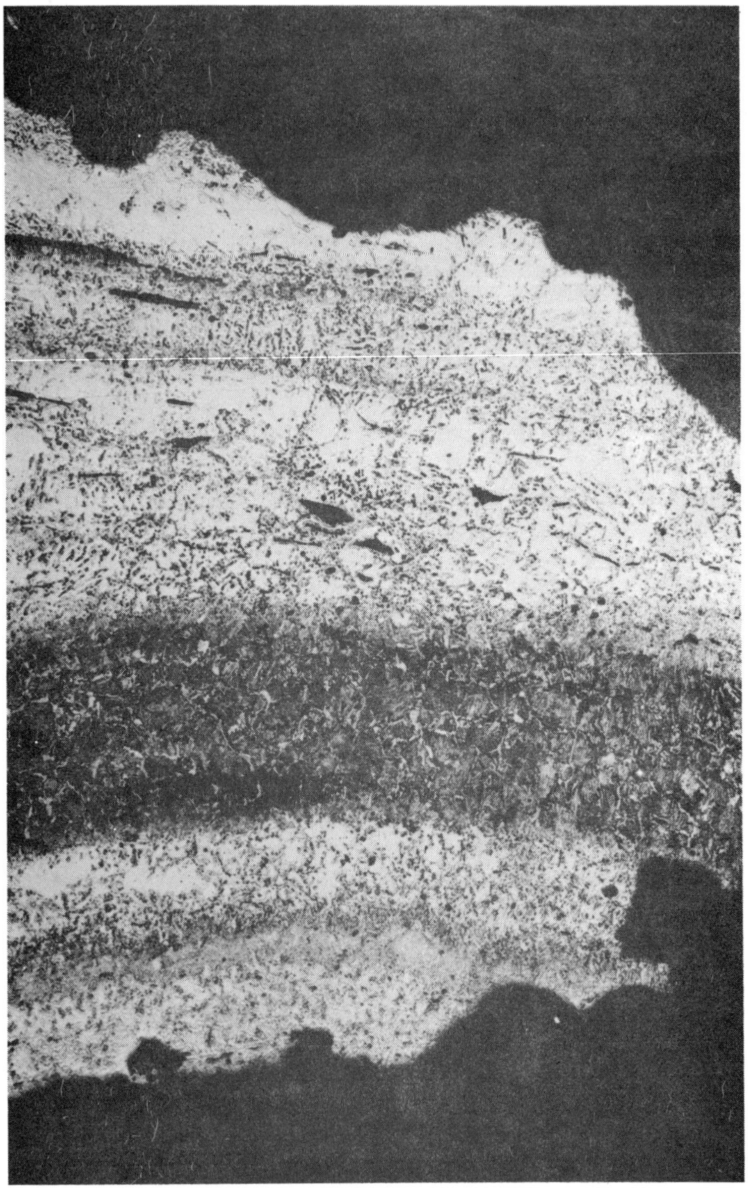

FIGURE 3.9

Iron with a central high carbon strip.
Wedge WD11a - 60X.

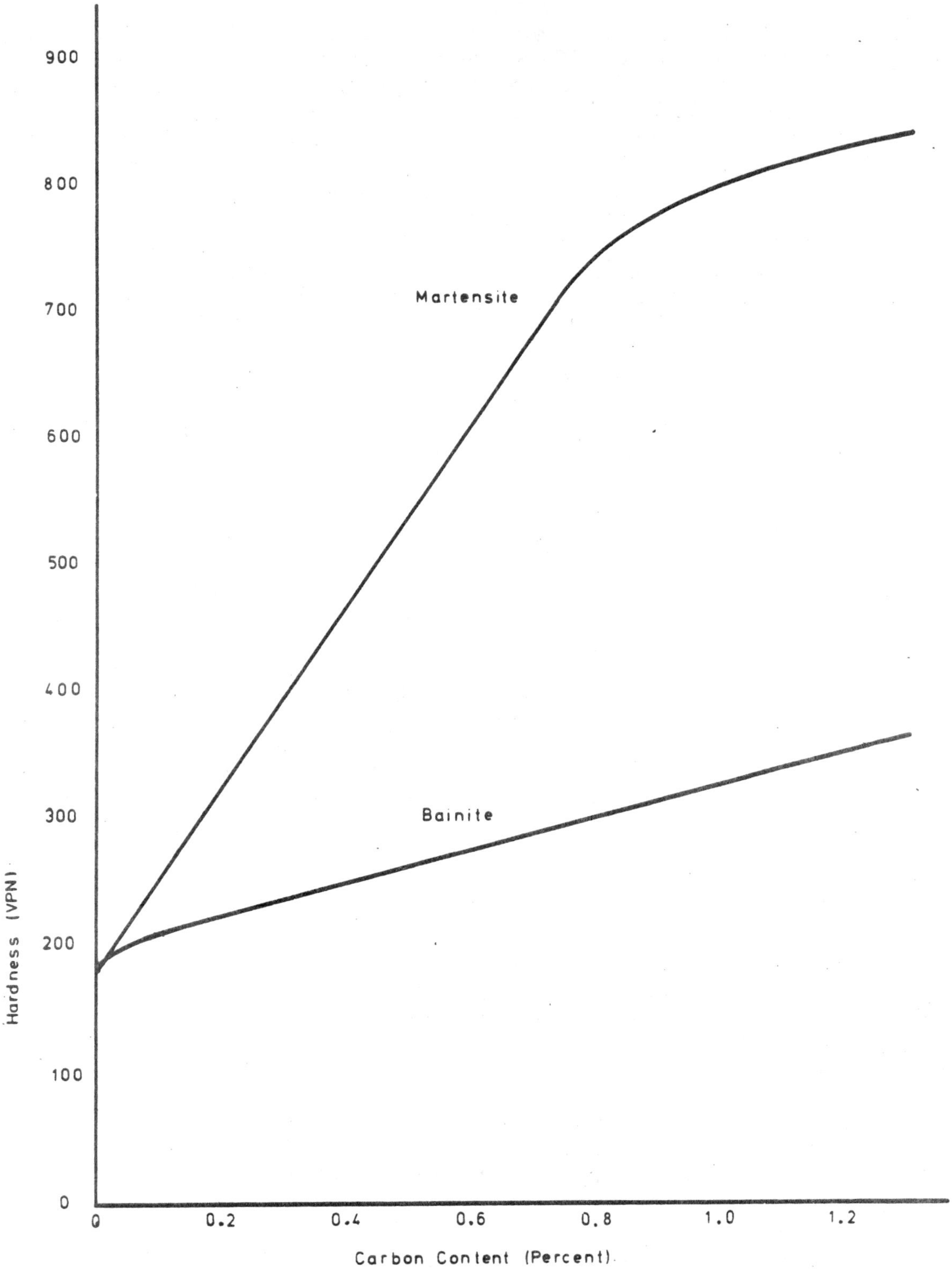

FIGURE 3.10

Graph of Carbon Content versus Hardness with the deliniations for bainite and martensite (Rollason, 1973, Fig. 128).

Phosphorus is a second impurity which affects the properties of iron. Phosphorus both hardens iron and makes it cold-short, or brittle at temperatures of 0°C and colder (Boylston, 1928, 446; Piaskowski, 1974, 237; Davies & Oelmann, 1983, 86). Phosphoritic iron does not etch well with a 0.2% Nital solution. A longer etching time is required before the grain structure will appear. As stated earlier, phosphorus is contained in bands, and prevents the diffusion of carbon in iron. Thus, on examination of an etched, high-phosphorus iron sample, alternating layers of ferrite, carburized iron, and high-phosphorus iron should be apparent. Phosphorus folations, or shadows, are also occasionally evident (Fig. 3.11). These are produced by the differential etching of the low-phosphorus and high-phosphorus iron intermediate zones. Although it is impossible to determine the concentration of phosphorus in iron by reflected optical microscopy, it is possible to note its presence.

A final grain structure of importance that can occasionally be observed under a reflected optical microscope is piling. Piling is the process of folding and welding strips of iron together to produce the body of an artefact. The strips should usually range from 5 to 10mm thick to enhance forgeability and weldability. Strips of iron which are too thin are difficult to weld since they cool very quickly from the required temperatures and burn very easily. Piling is usually performed to enhance the properties of a tool. Piling allows low-carbon and high-carbon iron strips to be easily combined to form the superior blade structure described earlier. Also, a large number of carburized iron strips may be piled to produce a more evenly distributed carbon content than possible by the carburization of a single thick iron item.

Piling can occasionally be recognized under a reflected optical microscope (Fig. 3.12). Each iron sheet may be differentiated by either thin lines of inclusions, or slightly darker etched streaks. No visual indication of piling in a sample, however, does not necessarily prove that the artefact was not piled. Samples HNY15a, HNY27a, and HNY52a were from artefacts which were split along one edge revealing separate sheets of iron. The samples taken adjacent to these splits, however, revealed no evidence of piling.

Thus, the reflected optical microscope was also employed to note any modifying criteria, such as piling, which may affect the properties of the artefacts.

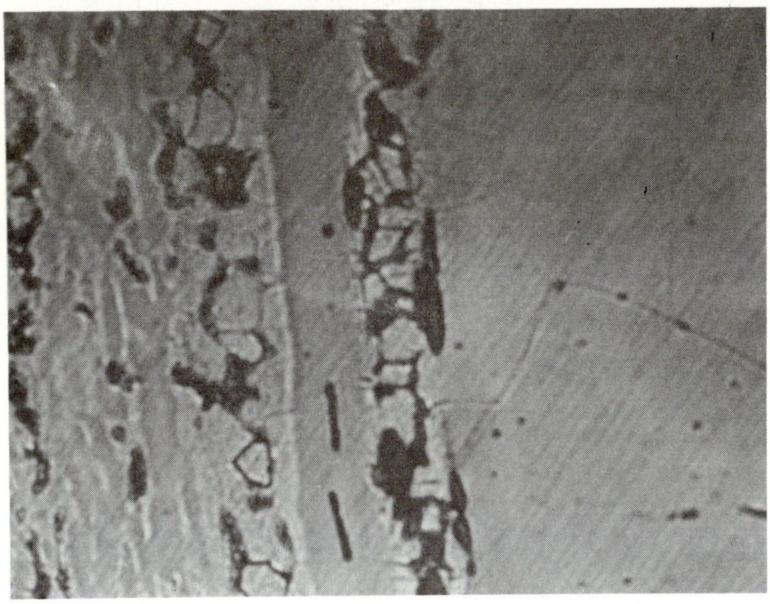

FIGURE 3.11

Phosphoritic foilations.
Spearhead BC4b - 350X.

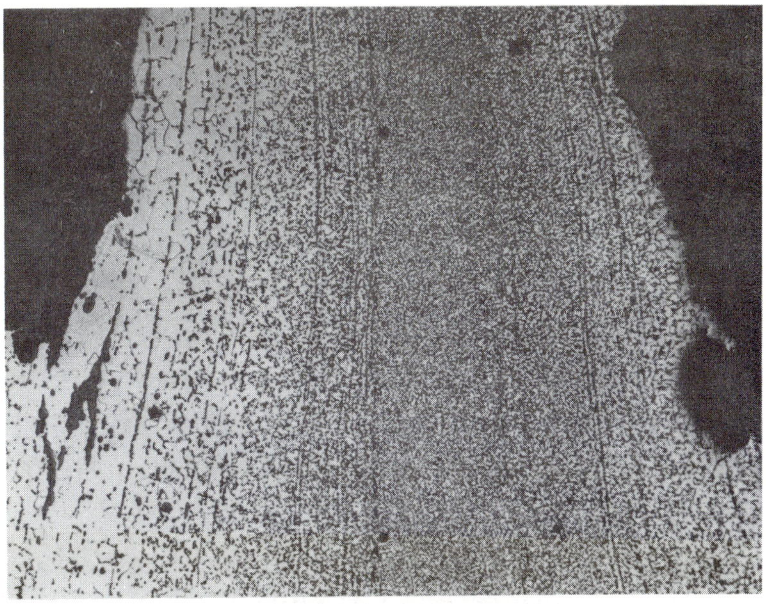

FIGURE 3.12

Piled iron.
Spearhead Dlb - 40X.

3.5.3 Trace Element Analysis

Trace element analysis using a Camebax Electron Microprobe allowed the determination of four important trace element concentrations: phosphorus, sulphur, cobalt, and nickel. Although the method by which the microprobe determines the trace element concentrations is beyond the scope of this publication, all of the relevant parameters required to reproduce my results are listed in Figure 3.13. The correction method used was the Standard Cameca ZAF x-ray correction program with a two-current modification added by Chris Salter. The only statistic which requires discussion is the Standard Deviation. Although this statistic is of extreme importance when attempting to distinguish between trace element concentrations of below 0.010%, the parameters of this study are sufficiently coarse that such discrepencies are negligible. Also, since the samples were all analyzed with the same equipment, the problem of inter-machine compatibility has been avoided.

The first two elements analyzed for were phosphorus and sulphur. As stated above, phosphorus increases the hardness of iron, but also makes it cold-short. Sulphur causes iron to be brittle at both room temperature and forging heat (Rollason, 1973, 169; Davies & Oelmann, 1983, 86). Iron with a sulphur concentration in excess of 0.035% would be impossible to forge, since the metal would fracture with every blow of the hammer. Thus, the microprobe results assisted in the determination of any extenuating criteria which would affect an iron artefact's properties.

The final two elements for which the artefacts were consistently analyzed were cobalt and nickel. Although neither of these elements affect the properties of iron, their presence is restricted to only one known ore source in Britain (see page 93). Salter's preliminary work (1984) on the Danebury iron tools revealed that the concentrations of cobalt and nickel fluctuated sufficiently to warrant further investigation of the extent of production and trade of iron manufactured by this ore source during the Iron Age. Thus, analysis for these elements continued.

Salter (1984) also analyzed the preliminary samples from Danebury for copper and arsenic contents. In the early stages of this study, the samples had not revealed sufficient fluctuations in these element concentrations to identify an additional ore source, and thus warrant further examination. Copper and arsenic also would not have greatly affected the properties of iron artefacts in the concentrations discovered, and therefore would have increased the research effort to little effect.

The trace element concentrations were determined for all completely corroded samples. Certain impurities, such as cobalt and nickel, present in the original metal remain

Lines examined for	Kα
Accelerating voltage	20 KV
Standardization current	10 nA
Analysis current	30 nA
Counting time	50 secs
Number analyses per sample	4
Co detection limit	0.009%
Ni detection limit	0.010%
P detection limit	0.001%
S detection limit	0.004%
Co standard deviation	0.002%
Ni standard deviation	0.002%
P standard deviation	0.002%
S standard deviation	0.001%

FIGURE 3.13

Parameters used for the microprobe analyses.

pinned to their locations with the corrosion of the iron artefact. Thus, their concentration may not substantially alter even after all the iron has rusted away. The danger, though, is that impurities can be absorbed from the soil. Phosphorus and sulphur can dramatically increase with the corrosion of an iron artefact. The only reliable manner for determining the original trace element concentration of a corroded artefact would be to test the soil surrounding the object also, so that all possible impurities, which might be collected by the artefact, can be allowed for. Unfortunately, this was not feasible in the present programme. Thus, the trace element concentrations of corroded samples, although noted, were omitted from the data employed in the formulation of the hypotheses presented in Chapter 4.

3.5.4 Vickers Hardness Testing

Vickers Hardness testing is a relational method for the determination of a material's hardness. A square-based pyramidal diamond point is forced into a sample under a selected load, producing an indentation. Depending on how hard the material is, the point will either be able to penetrate very deep, producing a large indentation, or very little, producing a small indentation. The extent of penetration is determined by measuring the two diagonals of the indentation's square base, and averaging them. This result can then be found on a Vickers Hardness table for the specific load, revealing a relational hardness for the material.

Vickers Hardness testing provided the only indication of the mechanical properties of the iron samples. Vickers Hardness testing assisted in the identification of quenched artefacts, and property anomalies indicative of either high concentrations of certain impurities or cold-working. Finally, testing permitted the determination of the suitability of certain iron alloys and tools for specific tasks.

3.6 The Chronological Framework

The Chronological Framework for the Danebury site is the most comprehensive so far compiled for Britain (Fig. 3.14; Cunliffe, 1984a, 234). It is based on a well-defined sequence of ceramic styles tested by both stratigraphy and various forms of seriation, and was designed to be applicable to Wessex. A series of 53 radiocarbon dates obtained from well associated contexts in Danebury allowed, after statistical manipulation (Cunliffe & Orton, 1984, 197), broad absolute date brackets to be assigned to the sequence (Fig. 3.15). The Danebury Chronology can be correlated with a coarser system of style-zones defined by Cunliffe some years ago (Cunliffe, 1966).

Each site from which artefacts were sampled was investigated for both the reliability of the excavation and the contexts of the artefacts. If associated pottery was discovered with the iron artefacts, Professor B. Cunliffe was kind enough to advise on their phasing in relation to the Ceramic Phase Chronology determined for the Danebury site (Cunliffe, 1984a, 234). The dating of the Northamptonshire sites was based on the chronology devised by Knight (1984, 95-99). If no associated pottery was discovered, or the artefacts were recovered in a haphazard manner, such as the quarrying of Hunsbury, then the artefacts could only be identified as Iron Age, Roman, or Saxon by their type with no further distinction possible. Although the lack of accurate dating did not eliminate these artefacts from general studies, they were of no assistance in the formulation of theories concerning the development of blacksmithing technology during the British Iron Age.

All 503 iron artefacts sampled were subjected to some of the three methods of analysis listed above. Sixty-one samples contained no existing iron, and therefore, could only be tested for trace element concentrations. An additional 73 samples contained too small a quantity of iron to be etched or tested with the Vickers Hardness tester. The remaining 369 samples were examined extensively.

Ceramic Phase	Diagnostic Characteristics	Style Zone Equivalent
1	angular bipartite bowls stamped decoration	Early All Cannings Cross
2	furrowed bowls	Later All Cannings Cross
3	scratched-cordoned bowl	All Cannings Cross-Meon Hill
4	plain round shouldered bowls	
5	simple barrel and jar forms	
6	plain saucepan pots	Yarnbury-Highfield and St Catherine's Hill-Worthy Down style
7	decorated saucepan pots	
8	early wheel-turned vessels and Dressel 1A amphorae	Atrebatic
9	developed wheel-turned	

FIGURE 3.14

The Danebury Chronological Framework
(Cunliffe, 1984, 234).

cp 1-3	550-450 BC
cp 4-5	450-400 BC
cp 6	400-300 BC
cp 7	300-100/50 BC

FIGURE 3.15

The date brackets for the
Danebury Chronological Framework
(Cunliffe, 1984, 197).

CHAPTER 4: THE ANALYSIS OF THE METALLURGICAL RESULTS

4.1 Introduction

Upon completion of the metallurgical analysis described in the previous chapter, the data were analyzed statistically using the Harvard University Peabody Museum's Vax 11/730 computer and the PHIND relational database system programmed by the author. The PHIND database system allowed manipulation of the 503 entries in a variety of ways not feasible manually. For instance, the selection of a specific tool type from the complete database and the sorting of these artefacts in a separate list first by age, and then within age by phosphorus concentration was a simple procedure. This rapid data manipulation allowed correlations to be performed quickly and simply, and permitted rapid and accurate insight into the factors relating to the production of iron stock and finished iron objects during the British Iron Age.

The data were analyzed in three parts. First, the results of the grain structure examination were manipulated to determine both the variations in the inclusion concentration per artefact type, and the extent of use of such hardening techniques as carburization, cold-working, and quenching. This research was conducted first for the entire assemblage, and then for each Ceramic Phase to reveal if any development in blacksmithing technology occurred during the Iron Age. The second section of analysis utilized the Camebax Electron Microprobe results to determine the possible use of phosphoritic iron for specific purposes due to its increased hardness. This section was also divided into a generalized account for all the Iron Age iron artefacts, and by Ceramic Phase. Finally, the variation in the sulphur concentrations of the artefacts was examined to determine if Iron Age blacksmiths could identify high-sulphur iron, and avoid it.

The results of the Vickers Hardness testing provided a measure of the mechanical properties of each artefact. This assisted in the formulation of hypotheses on the use of carburization, heat-treatment, cold-working, and trace element impurities by providing insight into the actual effects of each of these variables on the artefacts' properties.

4.2 The Results of the Reflected Optical Microscopy

The slag and forging inclusion densities were the first reflected optical microscopy results analyzed. Ten artefact types were initially selected to test the validity of the theory that the amount of time required for the production of an artefact is reflected by the quantity of inclusions retained. The number of artefacts discovered per inclusion concentration delineation for the ten artefact types examined is presented in Figure 4.1. The categories of spearheads, blades, sickles, daggers, and chisels were chosen since these artefacts would demand greater attention during production and thus should exhibit low inclusion concentrations. The inclusion densities for nails, rods, scrap (i.e. forging debris), and ingots were selected for comparison with the information recovered from the five previous categories. Nails, rods, scrap, and ingots are inconsequential artefacts, and should receive minimal work during their production. Thus, their quantity of inclusions should be higher than those of the artefacts requiring extensive forging.

The currency bars were also examined to assist in the determination of their actual purpose. If currency bars were unfinished swords requiring only final forging, then their slag and forging inclusion concentrations should be low. However, if currency bars were iron stock forged into an easily identifiable shape for consistency in trade, then the inclusion concentrations should be equivalent to those of the other rapidly produced iron artefacts.

As can be seen by the results in Figure 4.1, the spearheads, sickles, and chisels, as well as the nails and ingots are predominantly of low and medium inclusion densities. Also, the proportion of blades, daggers, rods, and scrap possessing low slag and forging inclusion concentrations is greater than that for the other categories examined. This unexpected similarity in the inclusion concentrations of high-quality and low-quality iron products reveals that the amount of time and attention invested per artefact type cannot be accurately determined by this investigation. Certain individual artefacts, like the Llyn Cerrig Bach tyre, may reveal indications of additional forging, but predominantly, the results could not be generalized for any particular artefact type.

The currency bars were the only artefacts which tended to be higher in slag and forging inclusion densities. Currency bars account for 23% of the high-inclusion material. Tentatively, these data suggest that the currency bars did not receive the additional treatment that their piled structures were previously thought to indicate. However, any conclusions based solely on the results of inclusion densities would be premature. Corroborative evidence has to be recovered from the currency bar's grain structure before definite theories can be formulated.

Evidence of cold-working was identified in 24 iron artefacts. The results have been tabulated in Figure 4.2. Predominantly, only single examples of cold-working were discovered per artefact type, suggesting that this process

Artefact Type	No. Low	No. Med	No. High
Spearheads	5	11	3
Blades	15	10	6
Sickles	14	15	7
Daggers	2	1	1
Chisels	5	6	1
Nails	9	9	6
Rods	10	5	4
Scrap	15	7	3
Ingots	1	2	0
Currency bars	6	15	16
Total	117	114	74

FIGURE 4.1

The number of artefacts per slag inclusion density for ten artefact types.

4 Nails	1 Axe	1 Hammer
6 Currency bars	1 Blade	1 Dagger
2 Rods	1 Spearhead	1 Saw
4 Sickles	1 Awl	1 Ring

FIGURE 4.2

A list of the cold-worked artefacts.

was not intentionally performed by Iron Age blacksmiths to enhance the properties of iron tools. Moreover, of the four categories producing multiple examples of cold-working, two are inconsequential artefacts (i.e. nails and rods). These artefacts were not annealed probably because they were not worth the time required. The data previously obtained from the examination of slag and forging inclusion densities would suggest that the currency bars also did not receive considerable attention during production; thus, the currency bars were probably not cold-worked to enhance their properties. The only semi-consistently cold-worked artefacts which would demand further investigation are the sickles. The four cold-worked, hooked blades originate from four different sites: one is a small sickle from Battlesbury Camp (Ceramic Phase 7); one is a large, tanged sickle from Wilsford Down; and the two remaining blades are large, socketed sickles from Cow Down (Ceramic Phase 1-3), and Winnal Down. The disjointed nature of the artefact morphologies, and the sites and dates of origin would suggest that these artefacts were not part of a consistently cold-worked artefact type, but individual cases either not meriting annealing during manufacturing, or work-hardened during use.

Unfortunately, the dating of the artefacts was not sufficiently accurate to allow analysis of the extent cold-working was performed per Ceramic Phase. Thus, no indication of the possible development of the use of cold-working to increase the hardness of artefacts during the Iron Age can be ascertained.

As stated in Chapter 3, the addition of carbon to iron increases the hardness of the metal, but also decreases its ductility and toughness. Thus, the use of high-carbon steel for a tool can be beneficial or detrimental depending upon the purpose for which the tool is manufactured. An artefact which requires the ability to be honed to a sharpened edge, and is resistant to wear, should be produced from high-carbon steel; an object which requires toughness and durability should not. Before a valid assessment of the ability of Iron Age blacksmiths to employ the proper steel for the correct purpose can be attempted, however, it must first be proven that Iron Age blacksmiths could actually produce or identify high-carbon iron.

Sorting the database by carbon concentration reveals that of the 329 samples yielding sufficient metal for analysis, 216 exhibited carbon concentrations of 0.1% or lower. An additional 73 objects contained carbon concentrations of between 0.1% and 0.5%; only 40 articles showed areas of higher carbon content. Although these data alone might imply that steel was unknown to the blacksmiths of the Iron Age, the possibility still exists that a few of the tool types may have been consistently, and thus deliberately, produced from high-carbon iron.

Figure 4.3 presents the 51 high-carbon artefacts discovered divided into their 14 respective categories by site. Also listed are the numbers found at each site of these same tool types, but of low carbon concentrations (i.e. under 0.3%). Some of the data are inconclusive. The number of swords, daggers, files, tyres, hammers, and wedges sampled is too low to indicate accurately whether these tools were deliberately carburized or not. Sufficient numbers of adzes, blades, chisels, large sickles, small sickles, ploughshares, spearheads, and punches or awls were sampled, however, to permit some tentative conclusions. Predominantly, the results found for the adzes, blades, large sickles, small sickles, ploughshares, spearheads, and punches or awls would support the theory that steel was generally unknown during the Iron Age. For the total number of artefacts sampled of these types, too few possessed high carbon concentrations. The chisels, however, show some possibility that a specific tool was consistently and deliberately produced from high-carbon iron. Ten of the fifteen chisels sampled possessed high carbon concentrations: six from Hunsbury, and four from Danebury. Three of the Danebury chisels date from Ceramic Phase 7. Also two of the chisels discovered at Danebury (D139 - Ceramic Phase 5 and D158 - Ceramic Phase 7) were quenched and tempered.

Only one additional tool of the 503 artefacts sampled was quenched and tempered: a wedge from Worthy Down (equivalent to Danebury Ceramic Phase 7). The Worthy Down wedge does not appear to have been quenched and tempered accidentally. The blade (Fig. 3.9) is composed of a high-carbon strip embedded in a low-carbon body. This is the finest configuration for a cutting edge. A blade forged solely from high-carbon steel is harder and sharper, but also more brittle. A more desirable carbon distribution for a blade is a layer of high-carbon steel situated between two layers of low-carbon iron. This allows the high-carbon steel blade to be honed to a sharp edge, yet be cushioned in a casing of more resilient iron. Such a carbon configuration can be manufactured in two ways: either a piece of steel is welded between two pieces of low-carbon iron, or an evenly distributed high-carbon steel is placed in a forge at a temperature of above 900°C with an oxidizing atmosphere for a shorter period of time than required to carburize the piece. This eliminates the carbon on the surface, leaving a high-carbon strip down the middle. The wedge was quenched and tempered to a hardness of 333 VPN. The chisels, the Llyn Cerrig Bach tyre, and the Worthy Down wedge suggest that an extremely restricted knowledge of high-carbon steel, quenching and tempering may have existed during the Iron Age.

	ADZES	BLADES	CHISELS	DAGGERS	FILES	HAMMERS	PLOUGH-SHARES	PUNCHES or AWLS	SICKLES, LARGE	SICKLES, SMALL	SPEARHEADS	SWORDS	TYRES	WEDGES
All Cannings Cross		● ○						● ○		●				
Barbury Castle		○						● ○	●●	●●	●●●			
Battlesbury Camp		●●●								●				
Battery Hill		○								●				
Bigbury		○	●	●		● ○○	● ○				●			
Cow Down									●					
Danebury	○	●●●●●●●●● ○○	● ○○○○		● ○	● ○	●	●	●●●● ○○○	●●●● ○○	●●			● ○
Fifield Bavant Down		●								●				
Hunsbury	●●●● ○○	●●●●●●● ○○○	●●● ○○○○○○	●● ○○	○	●● ○	●●●●●●● ○	●●●	●	● ○○	●●●●●● ○		● ○○	
Little Sombourne							●				●			
Old Down Farm		○○					●	●●	●●	●				
West Down, Chilbolten		● ○										○		
Wilsford Down									○					
Winnall Down									●					
Worthy Down		●					●	●●						○

FIGURE 4.3

Table of the number of tool types found per site with high and low carbon concentrations.

● Low carbon
○ High carbon

As stated in Chapter 1, the use of piling has been hailed by numerous researchers investigating prehistoric ironworking, as the method employed by Iron Age blacksmiths for increasing the carbon content of iron, and thus enhancing its properties. Piling is only a useful technique, however, if at least a portion of the individual sheets of iron from which the artefact was produced contain a carbon concentration in excess of 0.3%.

The grain structures of 17 artefacts, and the macro-examination of three artefacts exhibited indications of the use of piling. Suitable carbon concentrations for piling to enhance their properties, however, were evident in only seven objects: a spearhead (Ceramic Phase 6) and a strip (Ceramic Phase 7) from Danebury; a punch or awl (Ceramic Phase 7) from Worthy Down; a blade from Battery Hill Street; and a spearhead, a blade, and a fragment from Hunsbury. Again, the variation in the types, sites, and dates of these artefacts would imply that this technique was not deliberately performed by prehistoric blacksmiths to enhance the properties of these artefacts. Too few of the artefacts reveal sufficient carbon contents to state that the presence of carbon was anything but accidental. The previous conclusion stating that high-carbon steel was not generally understood by the Iron Age blacksmiths would also corroborate the theory that piling was not performed to alter an artefact's carbon concentration.

The artefacts which were piled, yet possessed insufficient carbon concentrations, are: three sickles (Ceramic Phase 9) and a spearhead (Ceramic Phase 9) from Bigbury; an object from Old Down Farm; two currency bars from Meon Hill; and three large ploughshares, a wedge, and a pick from Hunsbury. Two of the large ploughshares and the pick from Hunsbury revealed separate, split sheets along their length, but exhibited no evidence of piling in the adjacent sample taken. The grain structures of these three artefacts would suggest that more artefacts may have been produced by piling than are identifiable by reflected optical microscopy.

Although the consistency of some artefact types might imply that piling was deliberately employed in the production of certain tools, it must be remembered that the currency bars were also produced by welding sheets of low-carbon iron together. Thus, many of the artefacts may have possessed piled grain structures due to being manufactured from currency bars. We may conclude that piling was not employed to form either more even carbon contents in iron tools, or join low-carbon and high-carbon iron together to produce superior tools, but was performed for some, as of yet undetermined, reason.

4.3 The Results of the Trace Element Analysis

Phosphoritic iron has significantly different characteristics than pure iron, and the carbon-iron alloy. The presence of minute quantities of phosphorus both hardens iron, and makes it excessively cold-short, or brittle at $0^{o}C$ and colder (Boylston, 1928, 446; Piaskowski, 1974, 237; Davies & Oelmann, 1983, 86). The ASTM handbook and the Smithells Metals Reference Book both state that phosphorus concentrations exceeding 0.080% in iron are exceedingly detrimental, and impart no benefits to the metal under any circumstance (ASTM, 1983; Brandes, 1983). Metallurgy texts published circa 1910, however, debate the extent of the effects of phosphorus in iron. J.M. Camp and C.B. Francis, in the The Making, Shaping, and Treating of Steel (1920, 573), assert that phosphorus 'is another element that has been painted a little blacker, perhaps, than it should.' Camp and Francis conducted experiments on iron samples containing phosphorus concentrations of between 0.018% and 0.110%. Their conclusion was that high-phosphorus iron withstood cold working tests as well as low-phosphorus iron. They later warn of the detrimental effects of phosphorus, but still state that it is difficult to determine a definite concentration above which phosphoritic iron should be discarded.

F.W. Harbord (1904, 604) points out that in his experience phosphorus concentrations of up to 0.25% in iron showed no ill effects in rolling-mills. However, he writes: 'cold bending tests with a falling weight (not in a press) are probably among the most practical tests for high-phosphorus material, but here again we find the influence of phosphorus is very capricious, some extraordinarily good results being obtained with material which would be distinctly dangerous to say the least, for use in structural work' (Harbord, 1904, 605).

Karsten, however, (Handb. d. Eisenhuttenkunde, 1841; reiterated in: Percy, 1864, 64; Greenwood, 1884, 207) believed that concentrations of phosphorus in iron of up to 0.3% showed only an increase in hardness without any rise in cold-shortness. This iron, he states, is of the best quality. At 0.5% phosphorus, 'there is still little cause for apprehension respecting the quality of the iron' (Percy, 1864, 64). It is not until the phosphorus concentration reaches 0.75% and above that Karsten believed the metal became cold-short and not of general use (Greenwood, 1884, 207).

The discrepancies between the phosphorus concentrations considered permissible are considerable. One possible explanation for this deviation in theories is that the requirements for iron have slowly become more stringent since the Industrial Revolution. Iron and steel have had to withstand more abusive treatment within the last two centuries than ever before. Also, the failure of iron and steel parts had not previously led to such disastrous consequences. If a hand tool breaks, the amount of time required to fix it, make a new one or just switch tools is negligible compared to the disruption or catastrophe caused if one part of an engine, boiler or train-rail fractures. Therefore, what is considered seriously detrimental must be seen from the perspective of the period in question, and not by the standards of another era.

Generally speaking, the phosphorus concentration for prehistoric iron remains below the 0.5% mark which Karsten (1841, 420) still designated as good quality iron. Only 31 artefacts, 17 of which are currency bars possess phosphorus concentrations of over 0.5%. The average VPN hardnesses for low-phosphorus, high-carbon, and low-carbon artefacts are 206 VPN and 148 VPN respectively. The average hardness for the high-phosphorus, low-carbon samples is 181 VPN. The phosphorus concentration of 0.080% was selected as the boundary because iron within this limit still possesses certain properties considered beneficial by present standards. Thus, 0.080% phosphorus is the maximum concentration to be found in certain modern alloys.

In order to determine whether Iron Age blacksmiths intentionally employed or avoided phosphoritic iron due to its unique properties, the iron tools were sorted in a manner similar to Figure 4.3. In Figure 4.4, the adzes, blades, chisels, daggers, files, hammers, ploughshares, punches or awls, large sickles, small sickles, spearheads, swords, tyres and wedges were divided first by site, and then within site by phosphorus concentration. The first two artefacts which appear to have been consistently manufactured from high-phosphorus iron are the ploughshares, and possibly the adzes. The number of adzes sampled, however, number too few and are too restricted in site locations to state definitely that adzes were deliberately manufactured from phosphoritic iron. More adzes from other sites and other periods would have to be sampled and analyzed before any conclusions could be formulated.

The currency bars were also predominantly produced from phosphoritic iron. Of the 53 currency bars sampled, only six are low in phosphorus: five from Meon Hill, and one from Hunsbury.

If the socket widths of the currency bars are correlated with their phosphorus concentrations, a definite relationship is revealed. The smaller currency bars (i.e. socket widths measuring less than 23mm) are low in phosphorus. The currency bars measuring at least 35mm wide just above the socket all contain high phosphorus concentrations. The only sample which does not match this general trend is artefact HNY8a. The currency bar handle sampled for HNY8a, however, is extremely unusual, and will

	ADZES	BLADES	CHISELS	DAGGERS	FILES	HAMMERS	PLOUGH-SHARES	PUNCHES or AWLS	SICKLES, LARGE	SICKLES, SMALL	SPEARHEADS	SWORDS	TYRES	WEDGES
All Cannings Cross		● ○						●●		●				
Barbury Castle		●						●●	○○	●●	○○○			
Battlesbury Camp		● ○○								●				
Battery Hill		●								○				
Bigbury		○	○	○		● ○○	●●				○			
Cow Down									●					
Danebury	●	●●●●●●●●●●● ○○○	●●● ○○○		●●	○○	○	○	●● ○○○○○○	●●●●● ○	●●			●● ○
Fifield Bavant Down		○								●				
Hunsbury	●● ○○○○	●●●●●●●●● ○○○	●●●●●●● ○○	●●● ○	●	●●●	● ○○○○○○○	●●●	○	●● ○	●●●●● ○○		●●●	
Little Sombourne							○				●			
Old Down Farm		● ○					○	○○	○○	○				
West Down, Chilbolten		●●										●		
Wilsford Down									●					
Winnall Down									●					
Worthy Down		●					○	○○						○

FIGURE 4.4

Table of the number of tool types found per site with high and low phosphorus concentrations.
● Low phosphorus
○ High phosphorus

be examined more fully in Chapter 6. It is important to note that the smaller currency bars correspond with Allen's spit-shaped bars (Allen, 1967, 310), while the large currency bars are Allen's sword-shaped bars (*ibid.*, 308).

A similar division of artefact type versus phosphorus concentration is evident in the hooked, cutting blades. Forty hooked, socketed blades were sampled of which 32 produced intact iron. Sixteen of the 32 blades can be classified as large, hooked blades. The large blades have socket-widths measuring over 30mm (Figure 4.5A). The remaining 16 blades are smaller, with socket-widths of under 25mm (Figure 4.5B).

When the hooked blades are sorted by increasing phosphorus concentrations, they divide almost perfectly into their separate tool types. The phosphorus concentrations of eleven of the sixteen large sickles are in excess of 0.090%. By contrast, twelve of the sixteen small blades contained phosphorus concentrations of lower than 0.064%.

This division of small and large hooked blades into iron of low and high phosphorus concentrations is more impressively displayed by the assemblage of Barbury Castle, Wroughton. The Barbury Castle assemblage contains 24 iron artefacts (MacGregor & Simpson, 1963). Four hooked blades were sampled, all of which produced intact iron. The two small sickles had phosphorus concentrations of 0.003% and 0.005%. The two large sickles contained phosphorus concentrations of 0.165% and 0.463%.

Sorting the hooked cutting blades first by age, and then within age by phosphorus concentration reveals a further insight into the development of the production of these two tool types. All eleven of the high-phosphorus large, hooked blades, except one Ceramic Phase 3 blade could be dated to Ceramic Phases 6-8. Two of the five large, hooked blades with phosphorus concentrations of below 0.068% are dated to Ceramic Phase 9, and two are from Ceramic Phase 3. The remaining large, low-phosphorus hooked blade was discovered without any corresponding datable artefacts.

Eleven of the twelve small, low-phosphorus sickles were from Ceramic Phases 6-8, and one was dated to Ceramic Phase 3. Of the four small, high-phosphorus blades discovered, three are from Ceramic Phases 6-8, and the final small blade could not be assigned to any specific phase.

The data presented above indicate that during the earliest part of the Iron Age, both the small and the large, hooked blades were produced from low-phosphorus iron. By the start of Ceramic Phase 6, though, the large

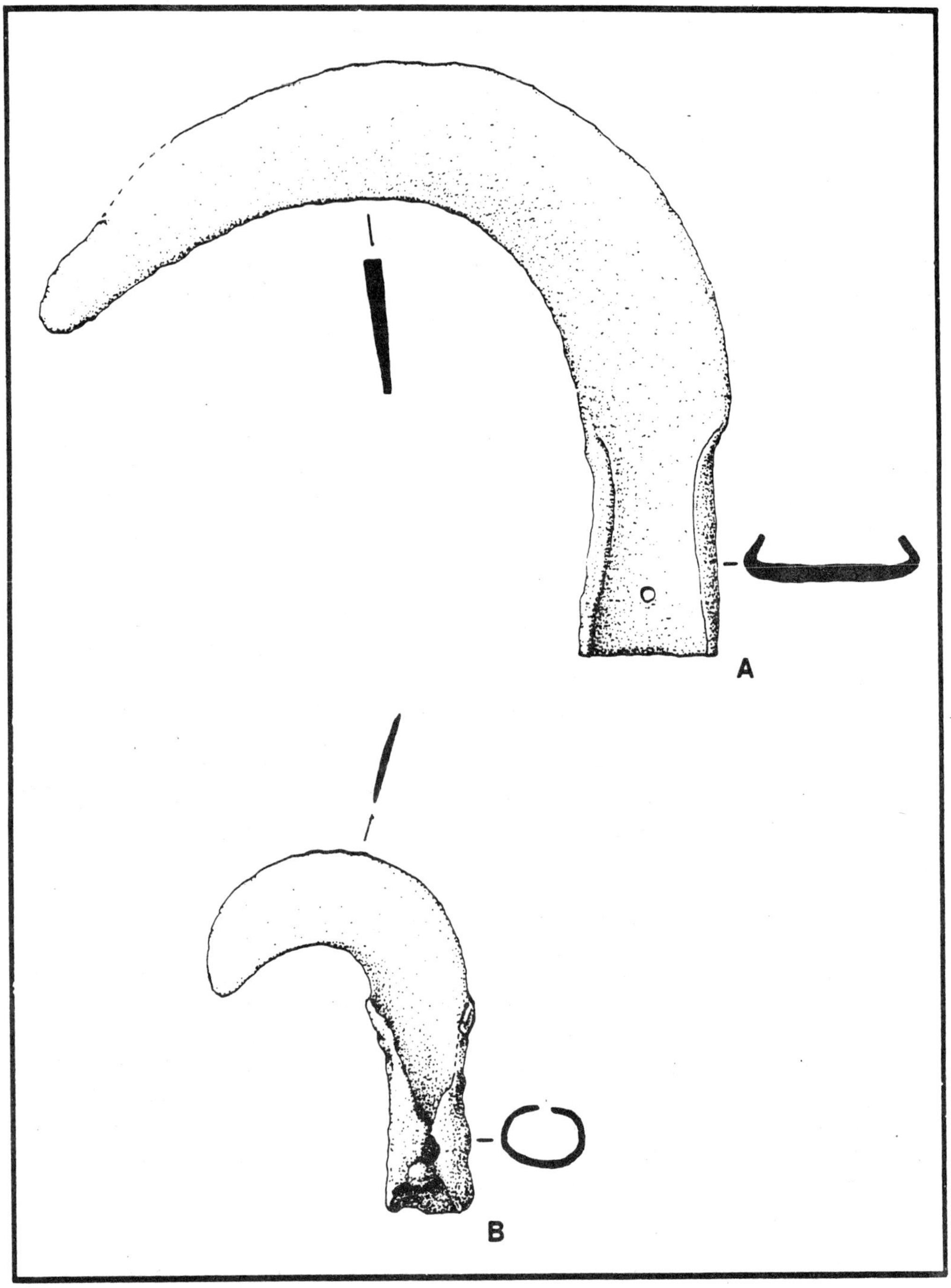

FIGURE 4.5

Two typical socketed, hooked blades.
A. Large blade - Danebury Small Finds 866.
B. Small blade - Danebury Small Finds 211.
Scale 2:3.

hooked blades were being manufactured from phosphoritic iron, while the small, hooked blades predominantly remained the product of low-phosphorus iron.

In order to determine whether this increase in phosphoritic iron was characteristic only of the hooked blades or reflected a general trend in the Iron Age, the entire database was sorted by phosphorus concentration within each Ceramic Phase. Of the 19 samples from iron artefacts dated to Ceramic Phases 1-3, only two possessed phosphorus concentrations of over 0.070%. By Ceramic Phase 7, however, the quantity of high-phosphorus iron artefacts had increased. Sixty-six of the 110 iron artefacts sampled from Ceramic Phase 7 possessed phosphorus concentrations of over 0.086%. Thus, either new sources of phosphoritic iron had been discovered by phase 6, or the altered properties of the metal had been discerned.

Sulphur is another detrimental impurity in iron. Concentrations of sulphur in excess of 0.035% cause the metal to be brittle, and fracture easily when forged (Rollason, 1973, 169; Davies & Oelmann, 1983, 86). All but three of the artefacts possessing intact iron revealed sulphur concentrations of below 0.035%. The first high-sulphur artefact examined was a spearhead from Barbury Castle, which possessed a sulphur concentration of only 0.037%. The other two artefacts possessing high sulphur concentrations are nails: one from Little Waltham, containing 0.060% sulphur, and the other from Bury Hill, containing 0.063% sulphur. The inconsequential nature of two of the three high-sulphur artefacts, with the third being only 0.002% over the limit, might suggest that Iron Age blacksmiths could not only identify high-sulphur iron, but also relegate it to minor uses.

CHAPTER 5: THE STATE OF IRON TECHNOLOGY IN IRON AGE BRITAIN

5.1 Introduction

This chapter will present the conclusions derived from the results of the metallurgical analysis described in the previous chapter. The sophistication of Iron Age blacksmithing technology will be assessed based on the studies of slag and forging inclusion concentration, cold-working, and piling, and the possible selected use of high-carbon steel, and phosphoritic iron. Evidence of the development and diffusion of techniques for the hardening of iron during the Iron Age will also be presented. This chapter will conclude with a summary of the state of iron technology in Britain during the Iron Age, and its implications for the organization of the blacksmithing community during this period.

5.2 The State of Iron Technology in the British Iron Age

The data recovered from the study of the grain structures of the 503 samples suggest that the Iron Age iron industry was predominantly unsophisticated. The quantity of slag and forging inclusions present per artefact type does not appear to have varied according to the importance of the object. For the low-quality nails and ingots to contain approximately the same inclusion concentrations as the high-quality blades and daggers indicates that the amount of preparation performed on the iron before being forged into finished artefacts was constant, and independent of the type of object to be produced.

The variations in the inclusion concentration of the artefacts sampled from Ceramic Phase 7 could be the result of the medium and high levels of inclusions contained in the currency bars. As stated in Chapter 2, after iron has been smelted, the bloom has to be cleaned and unified by heating and hammering it. This process has to be performed repeatedly before the iron will be useful. The iron for the currency bars was probably forged to a reasonably consistent quality first, and then formed into the rough sword shape. Generally speaking, a lower slag content in the currency bars was the result of either increased heating and hammering, or lower initial slag concentration. The high quantities of slag and forging inclusions found in most of the currency bars would have proved detrimental if the artefact was to be used as a sword. This amount would have been reduced to a reasonable quantity, though, if the iron was to be cut up and forged into finished artefacts. Depending on how dirty the original currency bar was, the final object would probably contain in the order of 1 to 2% less inclusions on completion of manufacturing. Thus, the inclusion concentration would not be the result of variation in the amount of attention paid to the production of an artefact, but the quantity of inclusions contained in the metal before production of the final artefact began.

Indications of the method for cleaning iron during the Iron Age should be difficult to identify, and irrelevant. In the case of the currency bars, however, many questions still remain concerning why piling was used in their production. It has been shown in the previous chapter that piling was not employed during the Iron Age to increase the carbon concentration of iron tools, since the existence of high-carbon steel appears to have been of restricted knowledge during this period. Why then should the time consuming process of thinning iron to sheets, and then welding these sheets together to form a bar be performed? One possible explanation is that this was the standard method by which iron was cleaned during the Iron Age. The metal could have been thinned to a thickness of approximately 5 to 10mm after the production of the iron bloom to eliminate a large percentage of the slag. Once sufficiently clean, the sheets were then folded and welded together to form the main body of the currency bar. The concentration of inclusions in the currency bars was still high, but not such that normal forging could not produce a usable finished tool. Thus, piling could have been performed during the Iron Age for a completely different purpose then is employed presently.

Although this discussion has centred on the Ceramic Phase 7 currency bars, it can be assumed that if limited techniques for the preparation of iron were standard during this later period, the level of technology for the previous Ceramic Phases must have been at least the same, if not cruder. Thus, the practice of cleaning iron consistently before the production of any artefact, and not varying the preparation for the more important tools, would imply that the blacksmithing technology had not reached the point of sophistication where methodology could be altered according to the planned function of the final product.

Techniques for hardening iron appear to have been generally unknown during the Iron Age. The four methods for enhancing a tool's properties which were considered in this study are cold-working, quenching, and the selection of the harder high-carbon and high-phosphorus iron alloys. Cold-working offers the versatility that it can be simply and accurately applied to any chosen area of an artefact without affecting the entire object. For example, a cutting edge of a knife can be cold-worked to enhance its longevity and sharpness, while the hammering of the rest of the blade can be avoided, thus leaving it softer and more durable. The paucity of cold-worked artefacts discovered dating from the Iron Age, however, would suggest that this technique was not exploited during this period.

The use of high-carbon iron to improve the properties of specific tools was also not regularly employed during the Iron Age. The majority of the artefacts sampled are low in carbon. The analysis of the chisels, however, indicates that a limited knowledge of steel may have existed. If the superior properties of high-carbon iron were understood by some blacksmiths, however, why should the tools predominantly be produced from low-carbon iron? This question is especially relevant for the sites of Danebury and Hunsbury, which yielded all of the carburized chisels, yet few other high-carbon tools.

One explanation could be that the technique of carburization was known to only a few blacksmiths. The process may have been discovered by certain blacksmithing families through generations of experimentation. Such a discovery would have been regarded as proprietary, and not shared with other members of the blacksmithing community. The history of blacksmithing is filled with secrets for the production of superior tools similarly discovered and safeguarded (Agricola, 1950; Biringuccio, 1942; Theophilus, 1847). Although many of these ancient processes may seem absurd in present-day contexts, some have been proven valid by modern experimentation. For example, Martin and Foley (1979) showed that the practice ascribed to Weyland the Smith of feeding iron fillings to ducks to increase the metal's hardness is metallurgically sound. Experiments consisting of feeding iron fillings to chickens, collecting the droppings, and then recovering the metal by burning off the dung revealed that this process increased the nitrogen content of the iron from 0.019% to 0.13% (Martin & Foley, 1979, 38). Since nitrogen is also a hardening agent, this process would increase the hardness of the metal. Thus, it is not inconceivable that a blacksmith during the Iron Age discovered a method similar in ingenuity to Weyland the Smith's, enabling him to increase the concentration of carbon in iron.

A second explanation for the restricted use of high-carbon iron is that instead of a blacksmith discovering carburization, a metalworker may have realized while cleaning an iron bloom that certain parts were harder than others. Piaskowski (1974, 325) states that certain sections of iron blooms contain higher quantities of carbon naturally due to the positioning of the tuyeres in the furnace. The section of a bloom situated the farthest from the tuyeres receives the lowest quantity of oxygen, increasing the quantity of carbon monoxide in that area of the furnace, and raising the amount of carbon available to be absorbed by the iron. Thus, steel can be accidentally produced during the smelting process. If a blacksmith realized that certain sections of the bloom were harder during forging, and reserved these areas for the cutting edges of such tools as chisels, then a superior tool would be produced. Again, such a discovery may have been regarded as proprietary, and not distributed to the rest of the metallurgical community. If either carburization or bloom selection was discovered during the Iron Age, and the technique was kept secret, high-carbon iron tools would then have been prestige products rarely to be discovered in the archaeological record.

It should also be noted that the chisels were found to contain predominantly medium slag and forging inclusion concentrations, and not low inclusion contents. This common inclusion concentration suggests that the skill of the blacksmiths producing these tools was not drastically superior to that previously encountered. If these blacksmiths were exceptionally advanced, a difference in the method of preparing the iron before manufacturing should also be expected. The fact that the manufacturers of the chisels did not seem to work the iron any differently than the rest of the blacksmithing community suggests that these smiths were not particularly more talented than the rest of the ironworkers, but possessed a single secret for the production of superior iron tools unrivaled by all others.

The restricted knowledge of carburization or bloom selection may also explain the lack of quenched and tempered artefacts. For quenching to be effective, the carbon content of iron must be in excess of 0.3%. Seventy percent of the artefacts sampled contained carbon concentrations of under 0.3%. Thus, even if some artefacts were cooled sufficiently rapidly for quenched structures to occur, the average low carbon concentration of the iron would not have permitted any of the heat-treated grain structures to be produced.

The final area of discussion about Iron Age iron technology concerns the use of phosphoritic-iron. The increase in high-phosphorus iron by Ceramic Phase 7 would suggest that a large iron production industry had grown up around a phosphorus rich iron ore source. However, was the rise of this iron industry due to the exploitation of phosphoritic-iron's increased hardness, even though cold-short, or was it merely the emergence of just another source of iron? As stated in Chapter 4, the average hardnesses for low-phosphorus low-carbon, low-phosphorus high-carbon, and high-phosphorus low-carbon iron are 148 VPN, 206 VPN, and 181 VPN respectively. These Vickers Hardness results reveal that high concentrations of phosphorus can harden iron to a point approximately midway between that possible for low-phosphorus low-carbon iron, and low-phosphorus high-carbon iron. For a society which did not understand the processes of carburization and quenching, the only readily available source of harder metal would have been phosphoritic-iron. With time, blacksmiths may have realized that the iron smelted from certain ore sources possessed special properties. Each source would then have been exploited for its unique characteristic, and the iron would have been traded as different quality stock. Thus, high-phosphorus iron may have been deliberately selected in cases where increased hardness was desired.

To determine if the increase in use of phosphoritic iron during the Iron Age was due to its exploitation as harder, superior stock, four constructions were devised which would be true if high-phosphorus iron was selected for its special properties. First, the tools selected to be produced from phosphoritic-iron should be those requiring increased hardnesses to enhance their function. Thus, a correlation of tool type versus phosphorus concentration should be revealed. Secondly, if phosphoritic-iron was seen as superior iron stock due to its higher hardness, prestige goods, such as weapons and chariot tyres, should also be produced from this metal. Thirdly, again, if phosphoritic-iron was believed to be superior iron stock, richer sites should possess greater quantities of it than poorer sites. Thus, a correlation of the level of the site type in the settlement hierarchy versus the percentage of high-phosphorus iron artefacts should be evident. Finally, the sites nearest to, or producing, phosphoritic-iron should possess a greater percentage of high-phosphorus iron tools.

The four artefact types which were predominantly produced from phosphoritic-iron were the adzes, large sickles, ploughshares, and currency bars. The number and find spots of the adzes sampled were not sufficiently varied to permit any conclusions to be drawn. It would appear that the adzes were consistently produced from high-phosphorus iron at Hunsbury. This could be an idiosyncrasy of this one site, however, due to some undeterminable bias in the archaeological record. Adzes from other sites and definable to definite Ceramic Phases have to be sampled before any conclusions can be formulated.

The ploughshares and large sickles were definitely being deliberately produced from phosphoritic-iron. The simplest explanation for the consistency of the alloy from which these two tool types were manufactured would be that each was being mass-produced at a particular source which received only a particular iron stock. This theory, however, demands an unlikely degree of specialization within the Iron Age blacksmithing community. Since the large hooked blades and the small hooked blades are discovered together on six sites, they must have served different purposes. A site that could produce one hooked blade size would also have manufactured the other very similar blade. Thus, the selection of high-phosphorus iron for the large hooked blades must have been for a reason other than a regional manufacturing peculiarity. However, no explanation for why they were divided into these two categories has been determined.

Although the large sickles would have benefited from the increased hardness of phosphoritic-iron, the ploughshares would not have been affected. The function of an iron ploughshare tip is to protect the wooden point of an ard from being eroded or fractured. A ploughshare also has to be sufficiently tough so as not to fracture if it strikes a stone. Thus, a softer, more durable iron alloy would prove superior to a harder type.

Finally, the currency bars should not be considered as tools, but as iron stock. The high slag concentrations, the indications of cold-working, and the presence of Widmanstatten grain structures in some of the currency bars sampled previously, would suggest that the currency bars were being produced in quantity. The currency bars were undoubtedly iron stock produced in a size and shape identifiable to the blacksmithing community. The fact that Allen's sword-shaped currency bars are phosphoritic, while the spit-shaped currency bars are low in phosphorus content would suggest that the shape of the bars were varied according to the source from which the iron was smelted. The artefacts sampled by Hedges and Salter (1979, 165) similarly divide into the categories of high-phosphorus sword-shaped (i.e. Danebury and Gretton), and low-phosphorus spit-shaped (i.e. Beckford). Iron Age blacksmiths could then differentiate between the iron of different sources, and perhaps, select a particular iron stock depending on the tool to be manufactured.

The tools which were consistently, and thus deliberately, produced from phosphoritic-iron do not appear to follow any rational pattern. It is not simply that all the tools with cutting edges, or all the striking tools of one period were made from one alloy while the rest were produced from another. Thus, the construction stating that the tools requiring higher hardness should be made from phosphoritic-iron, does not hold true.

Few prestige items were also produced from high-phosphorus iron. Only two daggers, one each from Hunsbury and Bigbury, contain a phosphorus concentration in excess of 0.080%. In the cases of the swords, daggers, and chariot tyres, however, it can be argued that high-phosphorus iron may have been intentionally avoided since it is cold-short. It is preferable to fight with a softer sword which is dull and bends easily, than to rely on a sharper sword prone to fracturing. A similar example of the deliberate neglect of a hardening technique dates from the 19th century in the United States. During the expansion period in United States history, some tools were intentionally not quenched and tempered because the density of the population was so low that opportunities to obtain new tools or repair old tools were rare. Therefore, it was safer to have tools that needed constant sharpening and were tough, than to have sharper, brittle tools the loss of which could cause severe hardship for the owner (Professor Manning, pers. comm.).

The intentional selection of the tougher, low-phosphorus iron for producing swords, daggers, and chariot tyres, however, would still indicate that an extensive knowledge of the properties of the different metal types was possessed by the Iron Age blacksmiths. The results of the first two constructions, thus, may suggest that instead of being seen as a superior stock due to its hardness, phosphoritic-iron may actually have been avoided because it was brittle.

The correlation of site type versus percentage of phosphoritic-iron artefacts reveals a further point of interest. This study was limited to only those sites from which more than two artefacts were sampled. The results have been tabulated in Figure 5.1. First, six of the 12 sites studied produced assemblages with more than half of the artefacts being phosphoritic. Although this would appear to suggest that most of the sites employed this metal, the quantity of scrap metal discovered per site is not necessarily a relevant indicator of the acceptance of a material. Only three of the sites (i.e. Barbury Castle, Old Down Farm, and Worthy Down) which could be dated to Ceramic Phase 7 possessed more phosphoritic-iron tools than low-phosphorus iron tools. Of the sites sampled, only one, Barbury Castle, is a hillfort. Battlesbury Camp, Bury Hill, Danebury, and Hunsbury all possessed more low-phosphorus iron tools than phosphoritic-iron tools. Thus, half of the small sites (excluding All Cannings Cross and West Down) which possessed samplable iron tools contained more high-phosphorus iron tools, while only a fifth of the large sites exhibited the same. This reverse in the distribution of high-phosphorus iron per settlement type would suggest that phosphoritic-iron was not being hoarded on the large sites, but was being predominantly worked at the smaller settlement sites. Thus, it may be that phosphoritic-iron was not exploited for its increased hardness, but was simply more readily available to all members of society. The low-phosphorus iron, however, was being used predominantly at the hillforts. Ceramic Phase 7 was selected for this section of the study because it is the phase in which large scale high-phosphorus iron production began during the Iron Age.

The only discrepancy in this theory is that the nails of Winnal Down are predominantly low-phosphorus iron. Nails are inconsequential objects, and do not usually warrant superior material since they will be lost to the metal supply once used. Nails, however, may have been

Site	%Tools	%Total
Barbury Castle	50%	56%
Battlesbury Camp	40%	40%
Bigbury	53%	48%
Bury Hill	0%	60%
Danebury	38%	51%
Hunsbury	34%	37%
All Cannings Cross	20%	20%
Little Waltham	0%	50%
Old Down Farm	88%	59%
West Down	0%	0%
Winnal Down	0%	36%
Worthy Down	75%	92%

FIGURE 5.1

The percentage of both tools and assemblages manufactured from phosphoritic iron for each site from which more than one artefact was sampled.

regarded as prestige items during the Iron Age, since their use shows that the owner has sufficient wealth that he can forfeit some of his iron stock to such minor uses as construction.

The proportion of phosphoritic-iron used at its own source is the final construction for determining the reason for the increase in the use of phosphoritic-iron during the Iron Age. Hunsbury is the only site sampled which is close to a high-phosphorus iron ore source. The quantity of high-phosphorus iron tools and scrap discovered at Hunsbury, as revealed in Figure 5.1, accounts for less than 50% of the total assemblage. However, since the Hunsbury assemblage was not excavated under controlled conditions, but salvaged during quarrying, the objects may reflect a considerable period of occupation, possibly longer than that during which phosphoritic-iron was being smelted. When the percentage of high-phosphorus iron tools and scrap for Hunsbury is compared with that for Danebury, however, it can be seen that Hunsbury contains less than Danebury, which is much further from a high-phosphorus iron ore source. Thus, even those hillforts situated near high-phosphorus iron ore sources seem to have preferred to import low-phosphorus iron from farther afield than to exploit their own supplies. This observation would imply that high-phosphorus iron was not being exploited for its increased hardness, but rather, was regarded as lower quality, more generally available stock.

The only site in which phosphoritic-iron appears to have been exploited for its increased hardness is Bigbury. Figure 5.2 presents the eleven artefact types discovered at Bigbury divided into the categories of low and high phosphorus contents. The three iron artefact types which were found to contain low phosphorus concentrations are the cauldron hooks, ploughshares, and bridle bits. These three types of artefacts should be low in phosphorus since their uses demand toughness as opposed to increased hardness. The properties of the blades, axes, picks, spearheads, daggers, and chisels would be improved by using phosphoritic-iron, since the increased hardness would allow a finer, more durable cutting edge to be ground on the blades.

The only artefact which exhibited a phosphorus concentration not suitable for its use was the slave chain. Like the cauldron hook and the bridle bit, a chain should be tougher as opposed to harder. The hammers and sickles were also divided between the two phosphorus concentration divisions. The slave chain, the hammers, and the sickles, however, could have been anomalies for reasons not distinguishable from the archaeological record, such as prestige value, or affordability.

The data provided by Bigbury would complete the trend of the use of phosphoritic-iron in Iron Age Britain. High-phosphorus iron was probably introduced into the blacksmithing community at approximately the end of Ceramic Phase 6. The blacksmiths were able to distinguish between the ore sources of the iron by the distinctive shapes the metals were traded in. The low-phosphorus iron was distributed as smaller currency bars, while the producers of phosphorus rich iron manufactured the larger currency bars. During Ceramic Phase 7, the new phosphoritic-iron was relegated to minor uses. Initially, the blacksmiths would not have known the new metal's properties, and thus, would have been wary about manufacturing important items from it. With time, however, the blacksmithing community must have realized the enhanced hardnesses offered by the higher phosphorus concentrations, and began producing those artefacts requiring higher hardnesses from this alloy. Thus, the development of the exploitation of phosphoritic-iron was a slow process.

Artefact Type	No. Low P	No. High P
Bridle Bits	2	0
Cauldron Hooks	1	0
Ploughshares	2	0
Sickles	5	4
Hammers	1	2
Axes	0	1
Blades	0	1
Chisels	0	1
Daggers	0	1
Slave Chains	0	1
Spearheads	0	1

FIGURE 5.2

Table of the number of artefacts of low and high phosphorus concentrations for Bigbury.

5.3 Summary

Iron technology was in a dynamic state throughout the Iron Age, with the blacksmithing industry of southern Britain being both innovative and segmented. During the early phases, almost all of the iron available was free of phosphorus. By Ceramic Phase 7, however, a phosphorus rich iron ore source had been discovered, and was supplying over half of the iron used. Predominantly, the blacksmiths of the Early and Middle Iron Ages were unable to alter the properties of the tools they produced. Innovative techniques for the production of finer iron, however, were being attempted during the same period. Certain blacksmithing families may have stumbled on methods for increasing the concentration of carbon in iron, thereby producing superior products. Clearly, such techniques would not have been disseminated within the blacksmithing community. Finally, by Ceramic Phase 9, the rest of the Iron Age blacksmiths were able to alter the properties of the tools they produced by selecting iron of different trace element impurities. Certain sources were exploited for their superior iron due to their natural richness in phosphorus. Distinctions between the properties of the metals would have been based on the varying shapes chosen by the iron producers by which to trade their metal.

Too much should not be made of either blacksmithing family distinctions, or the concept of ironworking hierarchies during the Iron Age, however. As Vannoccio Biringuccio (1942, 370) states in The Pirotechnia: 'When, finally, I consider what this act (blacksmithing) is, it seems to me that everything in all kinds depends only on experience, since these craftsmen are people without plan, and most of them are crude country people, and if they know how to do one thing they do not know how to do another.' The existence of an established hierarchy of blacksmiths in the Iron Age is undoubtedly a false model of the state of iron technology. A broad distribution of blacksmiths each with his own secrets and specialties is probably more accurate.

CHAPTER 6: THE INTRODUCTION AND DEVELOPMENT OF IRON TECHNOLOGY IN IRON AGE BRITAIN

6.1 Introduction

Although the conclusions presented in the previous chapter may suggest that the Iron Age iron industry was stagnant, it was actually in a highly dynamic state. This chapter will discuss the methods by which iron technology was introduced into Britain, and developed during the British Iron Age.

The first half of this chapter will examine the period of transition between the Bronze Age and the Iron Age. Possible sources from which iron technology may have been introduced will be presented.

The second half of this chapter will discuss the increase in the availability of iron during the Iron Age. Evidence of both the development of new iron production centres, and the locations of possible iron ore sources will be presented.

6.2 The Introduction of Iron in Britain

The method and effects of the introduction of iron in Britain have been discussed for more than 50 years. Childe's view, published in 1940, was that iron was brought to the British Isles by a wave of Continental immigrants in the 5th century BC (Childe, 1949, 194). The new metal quickly gained popularity due to the general abundance of its sources. The population of the country could finally produce their own metal and tools, freeing themselves from the confines of the elite who manipulated the distribution of bronze. This resulted in a democratization of society, with all inhabitants finally being able to obtain and use superior tools and weapons (Childe, 1942, 182-183).

Although the role of immigration has subsequently been questioned, the theory that iron was adopted because of the general availability of its ore sources is still the predominant theory. Champion (1975, 142) believed that iron-working expanded 'when the increasing demand for tools and weapons necessitated the exploitation of a more readily available material...the economies to be made through the greater ease of obtaining iron ore may have outweighed the disadvantages of an unfamiliar metal in an attempt to satisfy demand.' According to Champion, the demand for bronze eventually rose beyond the quantity which Bronze Age society could produce due to the limited availability of copper and tin. This resulted in a shortage of metal stock which could only be overcome by the discovery of a new, more generally available material, such as iron. Thus, iron would have been adopted not because of any innate, recognizable superiority, but because of a severe industrial shortage.

Bradley (1984, 126-127) agrees with the premise of Champion's theory described above, but places greater emphasis on the social implications that a shortage of bronze might have had. He asserts that during the Bronze Age, an elite class could have existed based on the control of long-distance trade, especially that of bronze. The votive deposition of bronze goods during the Bronze Age could, thus, have been a ritual destruction of wealth by an elite class to display its superiority and reaffirm its power. If the competition for such roles was great, the elite may have had to reenact these demonstrations with increasing regularity. The increase in the complexity of both the trade routes and the alliances needed to obtain bronze, however, would have made such a social system precarious and vulnerable to fluctuations in the production by the distant sources. Thus, according to Bradley, when the demand for bronze exceeded that satisfiable by the industry, the entire system would have collapsed.

Cunliffe (1984b, 31) forwards a simple, plausible explanation for the alteration in the structure of society in Britain from the 7th to the 5th century BC. He states that the disruption of the trade routes between Europe and Britain from the 7th to the 6th century BC was a singularly devastating event, and could have caused the destruction of the basis by which the Late Bronze Age elite sustained authority. Cunliffe states that the reasons for the collapse of the trade routes are unknown, however, it was European-wide, and beyond the control of the Wessex region. Since the elite's wealth and authority were engendered by the control of imports, the disappearance of the trade routes would have eliminated their source of power. The flurry of hoarding displayed at the end of the Late Bronze Age could be indicative of the confusion produced by the disruption of the prestige system. To compensate for the collapse of the trade routes, the elite of Iron Age Britain needed to develop a new system by which authority could be displayed: this was quite possibly the control of land. Evidence of the restructuring of the British land-holding system is evident in the flurry of hillfort building in the 6th century BC, culminating with the emergence of developed hillforts by the 4th century BC.

Although the theories of Champion, Bradley, and Cunliffe summarized above would go some way toward explaining the process of transition from the Bronze Age to the Iron Age, the end of the period of hoarding and the status of iron on introduction have not yet been analyzed. The introduction of iron in Britain must be seen against the crescendo of hoarding that occurred during the Late Bronze Age. By the 9th and 8th centuries BC, the quantity of bronze deposited in Britain either as hoards or votive deposits, was immense. During the 7th century BC, however, the number of bronze hoards and votive offerings decreased, until they finally disappeared at the end of the 7th century BC. The expansion of hoarding in the 9th and 8th centuries BC, indicative of a crisis in the bronze industry, coincides with the time that iron first penetrated Europe. Whether the introduction of iron in Europe also affected the bronze industry cannot be determined because of the paucity of evidence; however, the question of why iron was seldom hoarded or deposited votively still remains. Burgess (1979, 275-276) believed that this alteration in hoarding was due to iron's introduction by Continental Gundlingen sword producers resident in Britain. According to Burgess, this new material would have started an industrial explosion which would eventually have made bronze useless. Thus, the hoards would signify the 'wholesale dumping, early in the 7th century, when the bottom fell out of the bronze market' (Burgess, 1979, 275). The practice of hoarding then ceased for three reasons: first, iron was too common to require the protection of burial; secondly, iron was not as easily recycled as bronze; and finally, iron corrodes rapidly in the ground, making such storage unreasonable. Although Burgess' arguments seem plausible for the Late Bronze Age alone, they are difficult to sustain when the material evidence is compared with that of the Late Iron Age. During the Late Iron Age, hoarding reappears with the advent of the currency bar. If hoarding ceased during the transition from the Bronze Age to the Iron Age due to the general availability of iron, why should the hoarding then reappear during the Late Iron Age when iron was even more common? Clearly the situation was more complex.

A second point of relevance concerning the introduction of iron into Britain is why were the earliest iron objects discovered so far agricultural implements (i.e. the Llyn Fawr sickle, and a percentage of the socketed axes), and not weaponry or jewelry? The lack of personal ornaments could, of course, be due to the virtual absence of inhumation burials during this period, which are the usual means of survival of such material in the archaeological record. However, this still does not explain the paucity of complete iron weapons.

There are two plausible explanations for iron's usage on introduction. The first concerns the possible source from which iron was introduced. Cleere (1981, 184) states that the bowl furnaces dating from the Iron Age are primarily clustered around the south-west of Britain (Fig. 6.1). This technology is distinct from the later slag-tapping furnaces of eastern Britain, and might suggest that the A.1 furnace was introduced into south-western Britain

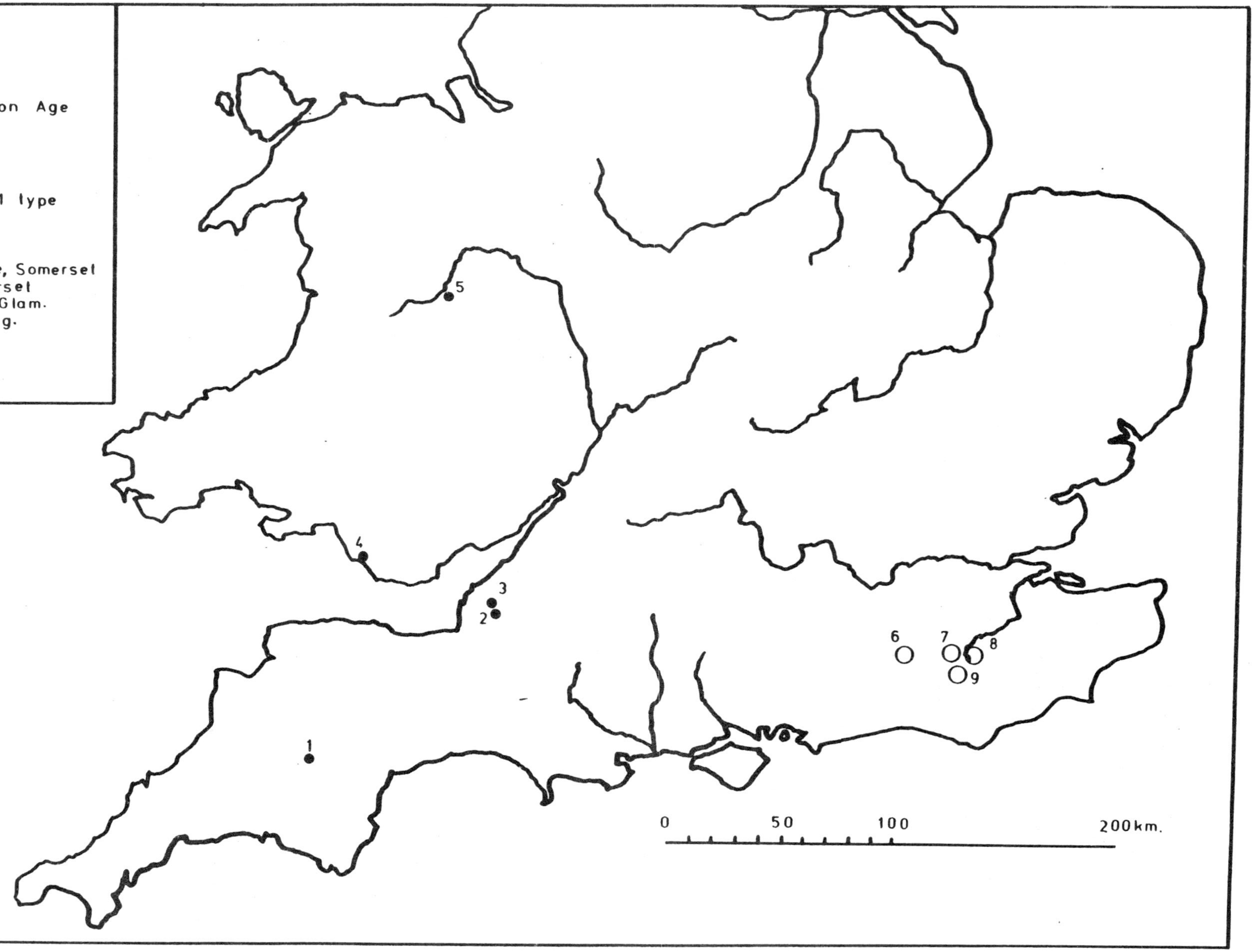

FIGURE 6.1

The distribution of Iron Age furnaces.

from north-west France. Unfortunately, the archaeological evidence of iron smelting furnaces in France is too meagre to indicate more accurately the location of the possible external origin for Britain's iron technology (Cleere, 1981, 184). The traditions of trade and contact along the Atlantic seaways, however, are well known (Rowlands, 1980, 37-41). Thus, it is not unreasonable to suppose that iron technology should have entered Britain along these routes. Also, the presence of the Llyn Fawr sickle, and the site of Trevelgue, the earliest large iron production centre discovered in Britain, would support the concept that iron was first smelted in the Devon-Cornwall region. This is also inherently reasonable since ironworking should have been introduced into an area of existing metallurgical significance. An experienced metalworking community would be more likely to be informed of metallurgical discoveries initially than a non-metallurgical community, and also be able to repeat the new smelting procedures successfully. Thus, the south-west of Britain could definitely be a reasonable point of entry for the new technology.

If the Devon-Cornwall region of Britain was the true point of origin of ironworking, and also possessed a flourishing indigenous bronze industry, it is an interesting speculation that iron may have been introduced initially in the form of low-status agricultural implements so as not to compete with the high-status bronze weapons.

A second possible explanation for the early use of iron for agricultural implements and its absence from hoards is that the new material may not have been part of the prestige system. The shift in the prestige system to the possession of land as the method of maintaining power may have been firmly established by the 6th century BC. Thus, metals may no longer have been significant.

6.3 The Availability of Iron

The extensive and meticulous excavation of Danebury has permitted valuable insight into the increase in availability of iron during the Iron Age. Danebury has produced the most accurate chronology so far available for the development of Iron Age ironworking in Britain. Of the 306 iron artefacts discovered at Danebury, 262 could be accurately assigned to specific Ceramic Phases. The fraction of the total iron assemblage found per phase is listed in Figure 6.2. The variation in the quantity of iron artefacts discovered per phase may reflect the growth of the iron industry in the Hampshire-Wiltshire region from c. 550 BC to c. 10 AD by revealing an increase in the quantity of iron lost or deposited through the Iron Age. This simple generalization is only valid, however, if the increase in iron on the site can be shown to be independent

cp	1-3	(550-450 BC)	-- 7%
cp	4-5	(450-400 BC)	-- 6%
cp	6	(400-300 BC)	-- 8%
cp	7	(300-100 BC)	-- 62%
cp	8	(100 BC - 100 AD)	-- 3%

FIGURE 6.2

The fraction of Danebury's total iron assemblage found per phase.

of such external factors as variations in the duration of the Ceramic Phases and fluctuations in the size of the resident community.

The different lengths of each phase may be compensated for by dividing the variable to be examined (i.e. the number of artefacts or weight of material per phase) by the duration of each phase in years. Sufficiently accurate dates for the eight phases of Danebury were available for this calculation to be successfully executed. If reliable population indicators are present on the site, the fluctuations in population may be adjusted for in the same manner: by dividing the number of artefacts or weight of material per phase by the number of people per phase. Unfortunately, Danebury did not produce suitable indicators of the size of each phase's resident community for this calculation to be performed. An alternative method for the determination of the effects of population changes on the quantity of iron discovered per phase is to compare iron's fluctuations with those of other artefacts of similar utility.

The graph of the total number of artefacts of iron, bronze, and bone per year per phase versus phase (Fig. 6.3) suggests that the factors controlling loss, or deposition, had the same general effect on all three materials. However, when the total weight of each material discovered per year per phase is plotted against phase (Fig. 6.4), the trend for iron deviates from those of bronze and bone. The quantity of iron lost increased with time, while the weights of bone and bronze rose and fell in the same manner as depicted in Figure 6.3.

Another deviation of interest in Figures 6.3 and 6.4 is the large decrease in the quantity of bronze discovered between phases 1-3 and phases 4-5. This decline is more dramatically revealed in the graph of tools per year per phase versus phase (Fig. 6.5).

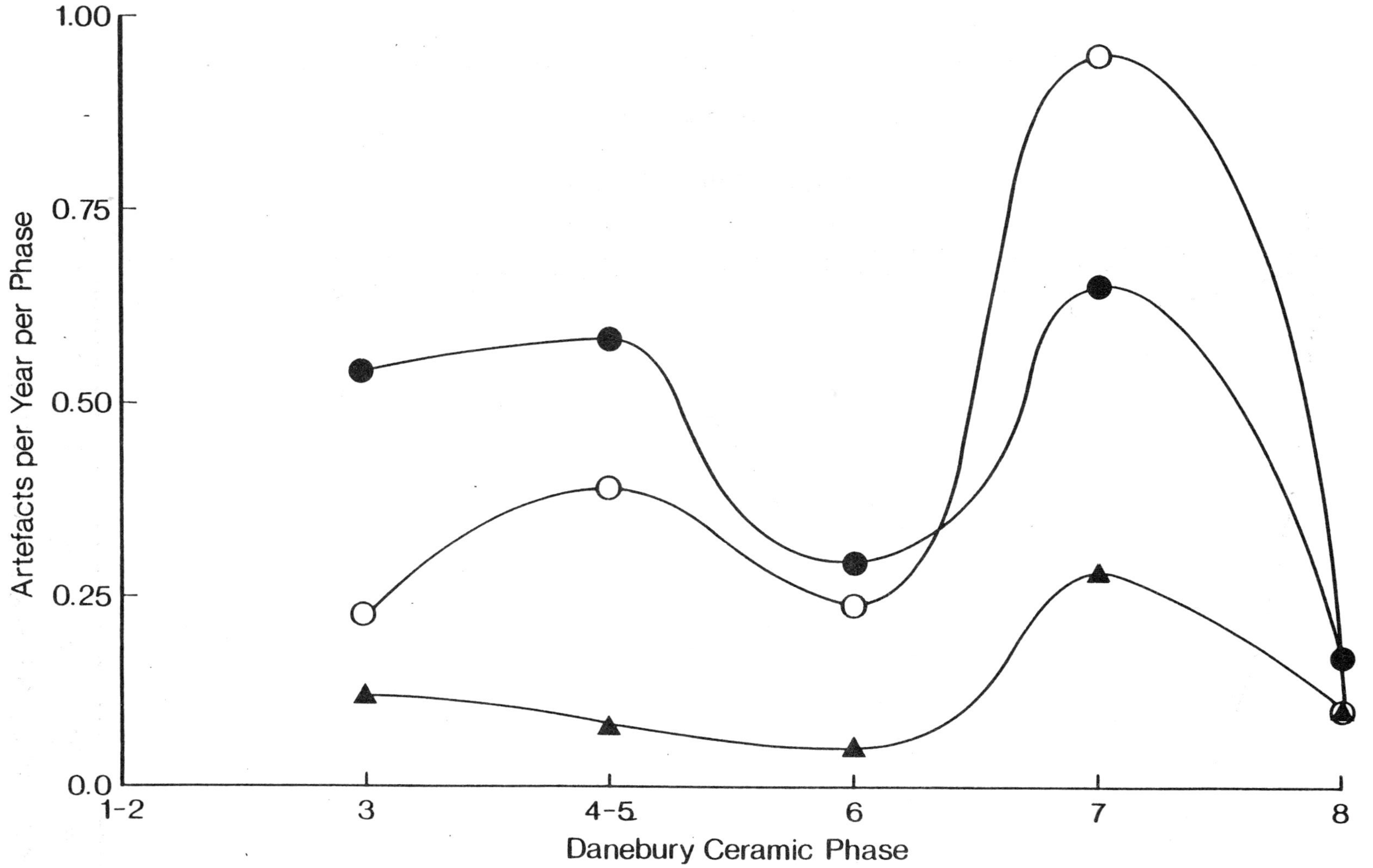

FIGURE 6.3

Graph of the total number of artefacts per year per phase versus phase.

○ Iron • Bone ▲ Bronze

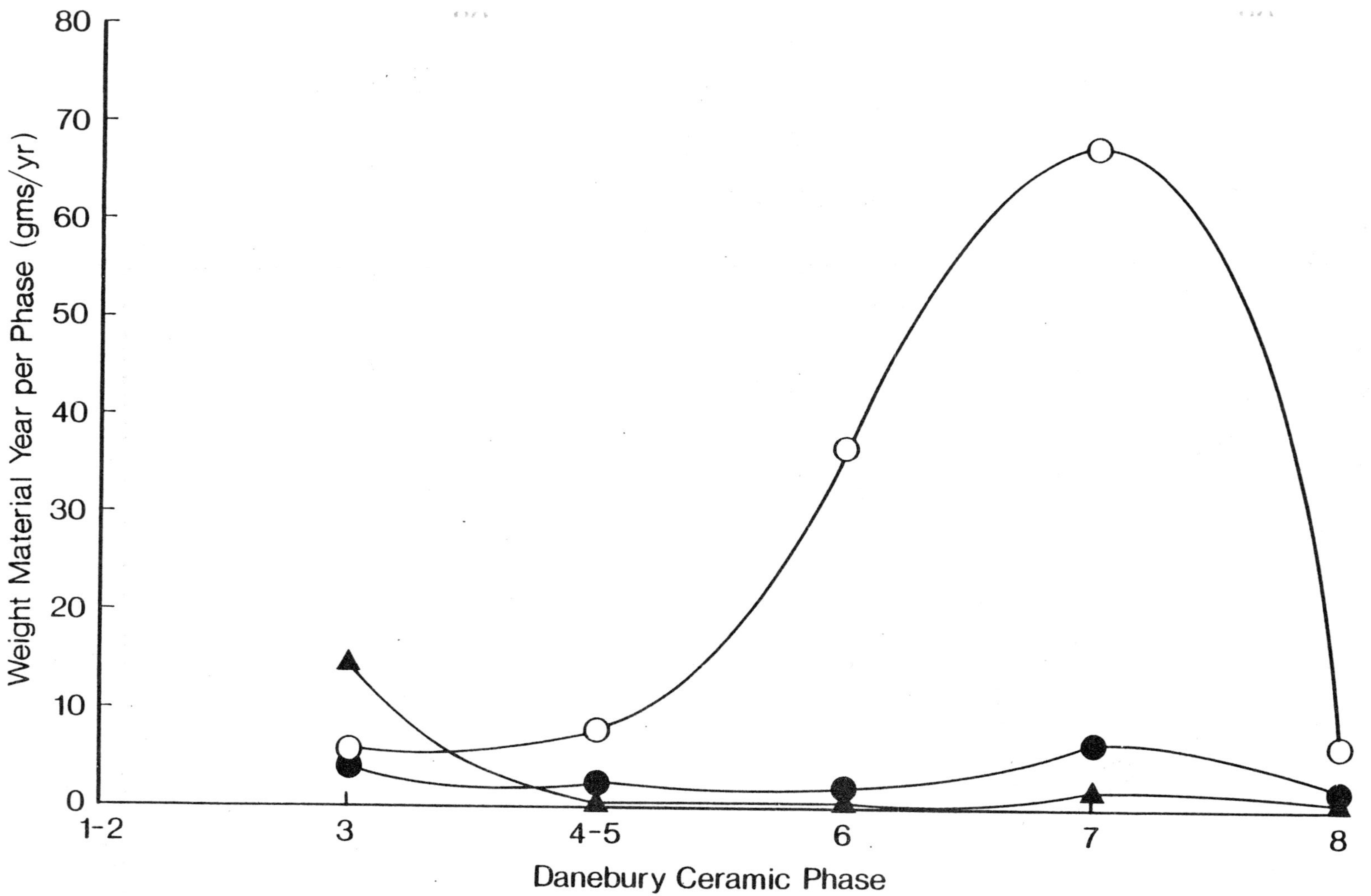

FIGURE 6.4

Graph of the total weight of material per year per phase versus phase.

○ Iron ● Bone ▲ Bronze

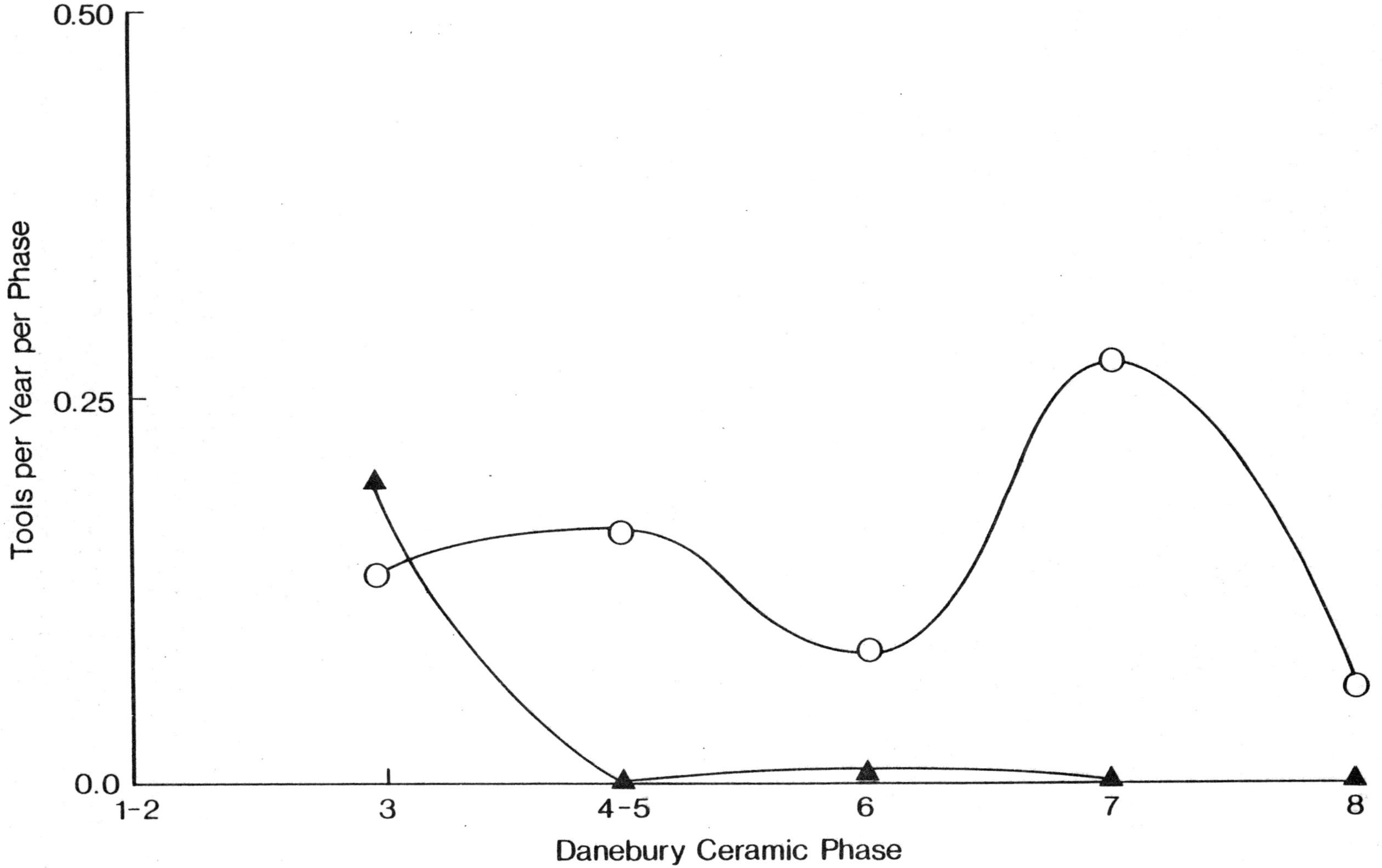

FIGURE 6.5

Graph of the total number of tools per year per phase versus phase.

○ Iron ▲ Bronze

The data depicted in Figures 6.4 and 6.5 indicate that with iron's increase during the Iron Age, the number of bronze tools decreased. Thus, the expansion of Iron Age iron manufacturing is likely to have diminished the extensive Bronze Age bronze industry. Although the verification of iron and bronze's cause and effect relationship is important, the results are not startling. The fall of the bronze industry with the rise of iron manufacturing has been thoroughly deliberated over the past 50 years (Childe, 1949, 178; Hawkes & Hawkes, 1953, 98-99). The sequential development of the Iron Age iron industry, however, has never been resolvable within the archaeological record beyond these two simple opposing trends.

It would appear that the remains of the large Iron Age iron production sites have been destroyed by subsequent smelting centres. Thus, it has been impossible to determine from present archaeological remains the origin and the method of expansion of iron technology in Britain during the Iron Age. Due to the lack of tangible evidence, archaeologists have been forced to interpolate the probable locations and development of the major iron production sites from both later areas of iron smelting and clusters of artefact types. The theories formulated attempting to define the development of the Iron Age iron industry have, unavoidably, been generalized and incomplete.

In an attempt to compensate for the lack of evidence of iron production sites in the archaeological record, alternative methods have been devised for determining the sources of iron smelted, the principle of which is trace element analysis. Every ore source possesses a characteristic set of trace elements which will be evident in all iron manufactured from it. Thus, in principle, the origin of an iron artefact can be determined by comparing its trace element concentrations with those characteristic of the known ore deposits. The validity of trace element analysis, however, is questionable. First, iron is not a homogeneous material. The trace element concentrations of one small sample may not be representative of an entire artefact. Secondly, ore sources do not possess consistent trace element contents throughout their bodies. Ore mined now from the lower levels of a source may not contain the same trace element concentrations as that collected in earlier times from the top levels. Finally, the ore sources exploited during the British Iron Age may have been exhausted. Any metal with trace element concentrations characteristic of depleted ore sources would then be either untraceable, or attributed incorrectly to another source.

Although the limitations of trace element analysis demand discretion when formulating theories, some useful information can still be gleaned from the data by reducing the ambitions of the study. The fluctuations in trace element concentrations per artefact can be compensated for by restricting the number of delineators per trace element to one only. For instance, the threat of mislabeling due to minor variations in the phosphorus concentration per artefact is reduced if the 0.080% concentration previously selected as the limit of the low-phosphorus category is also used for distinguishing iron produced from low-phosphorus and high-phosphorus iron ore sources. Thus, all that trace element analysis would indicate is whether the iron did or did not come from a high-trace element concentration source.

The only disadvantage with coarsening the conditions of trace element analysis is that it drastically increases the number of possible locations from which an iron ore could originate. Instead of being able to limit an iron type to one or two sources, the possible origin of the metal of an artefact is increased to entire regions, such as the Jurassic Ridge. The correlation of the trace element analysis results with the meager archaeological evidence, however, may permit some insight into the development of the iron industry during the Iron Age.

Trace element analysis revealed four possible iron types: iron with concentrations of cobalt and nickel in excess of 0.1%; iron with sulphur contents of over 0.035%; iron with phosphorus concentrations exceeding 0.080%; and the iron with trace element concentrations below those listed above. The artefacts with intact iron were separated using the PHIND computer system into these four classifications by concentration of cobalt and nickel first, then by sulphur content, and finally by phosphorus concentration. This method was designed to select the metals in the order of decreasing rarity of ore sources. Thus, a rare high-cobalt, high-nickel iron artefact also possessing a relatively common phosphorus content would appear only in the high-cobalt, high-nickel category. Each classification was then sorted separately by Ceramic Phase. The results are exhibited in Figure 6.6.

Plotting the percentages of each iron type per total number of artefacts per phase versus phase reveals how the different ore sources were exploited during the Iron Age. Pyrite and marcasite appear to have been smelted only sporadically. The high-cobalt, high-nickel iron decreased in use during the Iron Age, as did the low-impurity iron. The phosphoritic iron, however, gained favour through the period. There are two possible explanations for the decrease in use of both the low-impurity iron and the high-cobalt, high-nickel iron with the increase in use of phosphoritic iron: either the sites manufacturing the high-cobalt, high-nickel iron and the low-impurity iron were reducing production, thus forcing dependency on phosphoritic iron; or, the high-phosphorus iron source was increasing demand due to its being more generally available (it has been shown in Chapter 5 that phosphoritic iron was

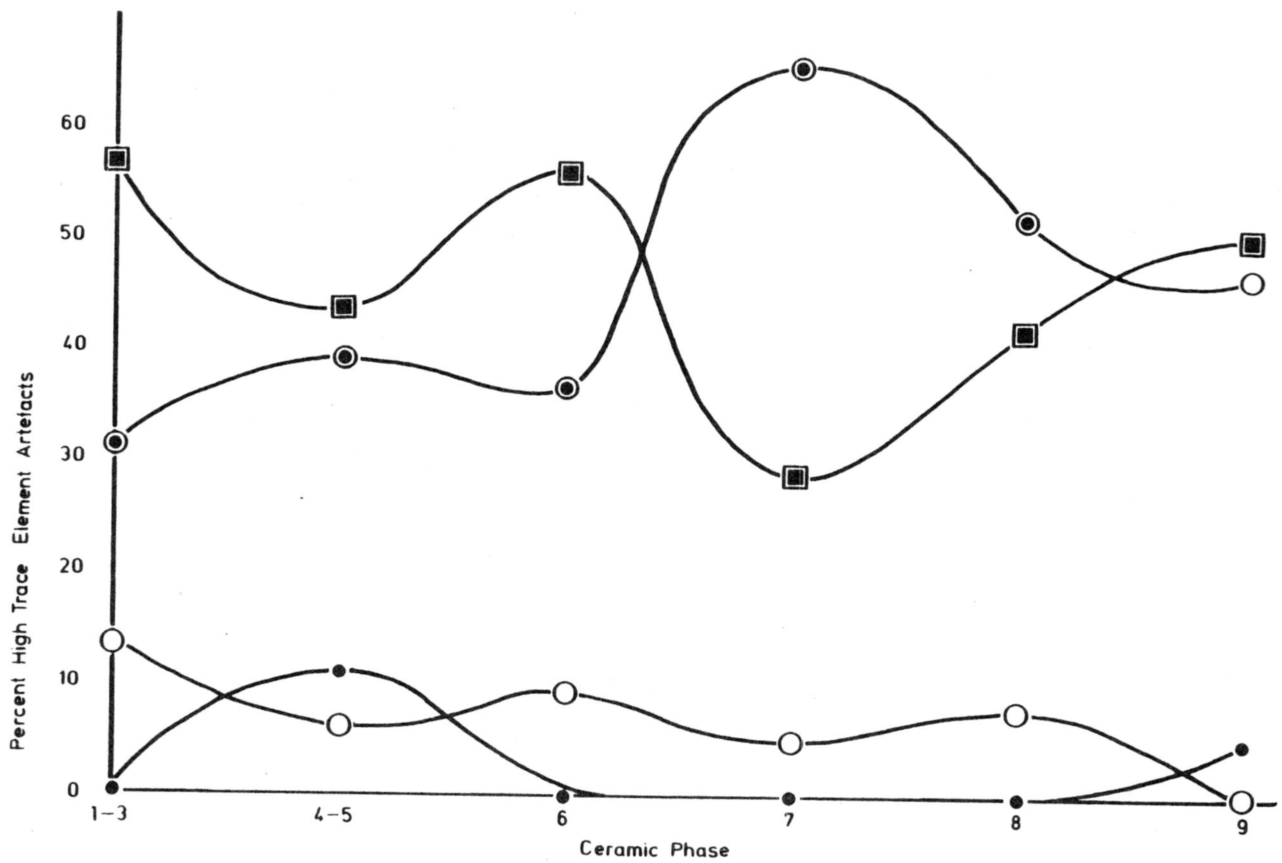

FIGURE 6.6

Graph of the percent of high trace element artefacts per Ceramic Phase

- • High sulphur
- ○ High cobalt nickel
- ◉ High phosphorus
- ▣ Low impurity

originally not selected for its higher hardness). If a decrease in the production of low-impurity iron and the high-cobalt, high-nickel iron did occur, it was not due to limited resources. Groves lists sources of both these ore types in his 1952 publication, showing that these ore sources were sufficiently large not to have been exhausted by the 20th century. It is also improbable that two iron production centres would fail simultaneously. Thus, the gain in favour of the phosphoritic iron is most likely due to its having been more readily available.

Proof that the rise in use of phosphoritic iron was the result of its general availability, and not to any reduction in the quantity of the other iron types produced may be sought by correlating the results of the trace element analysis with the meager Iron Age archaeological evidence. Iron low in all four of the trace elements would be impossible to trace to a specific source. Ores free of these impurities can be found in the Weald, the Forest of Dean, and the Lower Greensands (Groves, 1952, Tables 20 & 21; Salter, pers. comm.). Evidence of iron smelting for each of these areas, however, is limited. There is no convincing proof that the Weald was the source of the low-impurity iron; and Cleere's belief (1981, 184) that the Forest of Dean was the source of the currency bars, while possible, cannot be substantiated. Camerton is the only site to contain a sufficient quantity of slag to suggest large scale, low-impurity iron production (Cleere, 1981, 184). Camerton, however, was also a site of extensive Late Iron Age and Romano-British occupation. It should also be noted that the trace element concentrations of the European ingots discovered at the Isle of Portland were also low. Thus, this iron could even have been imported from the Continent. The limited nature of the information available from the archaeological record cannot account for a decrease in the quantity of low-impurity iron used during the Iron Age.

The sources of high-cobalt, high-nickel iron ore are much rarer in Britain. The only presently located source producing iron ore with high cobalt and nickel concentrations is near Newquay, Cornwall (Groves, 1952, Table 2). Cornwall has produced three iron smelting sites, of which only Trevelgue contained a sufficient quantity of slag to indicate large scale iron production (Salter & Ehrenreich, 1984, 151; Woolf, pers. comm.). Trevelgue is located near Newquay, very near the Great Perran lode (Cleere, 1981, 56). Further research has to be performed, however, to determine if the ore source exploited by the population of Trevelgue possessed high cobalt and nickel concentrations, supposing some part of it is still extant.

Although the link between the high-cobalt, high-nickel iron ore and Trevelgue is tenuous, the fact that the Devon-Cornwall region might have been the origin of ironworking in Britain may add to our understanding of the the high-cobalt, high-nickel iron source, and the place of this iron type in Iron Age society. Sites like Trevelgue could well be among the original iron-producing centres for Britain, and could have produced much, if not most, of Britain's iron in the Early Iron Age. The decline in the high-cobalt, high-nickel iron observed at Danebury might then have been the result of the later discovery of iron production sites within, or closer to, the Wessex region, making long distance trade with Cornwall expensive, and inefficient. This hypothesis is supported by both the distribution of the high-cobalt, high-nickel iron artefacts, and the evidence of iron production in the Wessex region during the Early Iron Age.

The known distribution of the high-cobalt, high-nickel iron is so far confined to three sites: Danebury (9 artefacts), Old Down Farm (1 artefact), and Winnal Down (1 artefact). Possible single artefacts also discovered are a nail from Glastonbury (Haldane, 1970, Table 2) and the Standlake sword (Salter & Ehrenreich, 1984, 152). The artefacts from Old Down Farm and Winnal Down could not be accurately dated. These, albeit very incomplete data for the distribution of the high-cobalt, high-nickel iron in Wessex could imply that this iron was imported into Danebury (and other hillforts), and occasionally redistributed from the fort to smaller neighbouring sites.

The importance of imported iron during the Early Iron Age is indicated by the smelting remains of the Wessex region. Tylecote (1962, Table 70) lists only three sites as possessing evidence of smelting for the Hampshire-Wiltshire region: All Cannings Cross, Little Woodbury, and Swallowcliffe Down. All three of these sites were occupied during the Early and Middle Iron Age, and the quantity of slag discovered on each would indicate that only limited iron production was performed to satisfy the site's individual requirements (Cleere, 1981, 184). The only site possessing similar smelting evidence dating from the Late Iron Age is Gussage All Saints. Gussage All Saints, however, appears to have been a specialized metallurgical establishment, producing bronze and iron horse trappings for trade, and smelting iron for its supplies (Spratling, <u>et al</u>., 1976; Wainwright, 1977; Wainwright, 1979). The fact that fewer Late Iron Age settlement sites seem to have been producing their own iron than Early and Middle Iron Age sites suggests that iron's availability was increasing. The inhabitants of central, southern Britain no longer needed to manufacture their own iron or import iron from great distances, but could obtain their metal stocks directly from a local, large iron producing site. Thus, the reduction in high-cobalt, high-nickel iron usage during the Iron Age could have been due to a rise in the availability of locally produced iron, and not to a decrease in the output of the high-cobalt, high-nickel iron production centres.

The iron artefacts rich in sulphur number too few to permit a valid estimation of the extent that pyrite and marcasite were smelted during the British Iron Age. The existence of only three high-sulphur artefacts, two of which were nails, would imply that the smelting of iron sulphide ore was occasional and the iron easily identifiable. Any sulphuric-iron that was produced was manufactured into inconsequential objects such as nails. The lack of high-sulphur iron, however, does not prove that pyrite and marcasite were not smelted during the Iron Age. Iron sulphide nodules can be purified by roasting (Tylecote, 1962, 199; Salter & Ehrenreich, 1984, 147). Thus, greater quantities of iron ore high in sulphur could have been smelted; it was only the iron resulting from partially roasted ore that would betray its origin.

As stated in Chapter 2, phosphoritic iron is predominantly discovered in two geological formations: the Jurassic Ridge, and the Coal Measures (Groves, 1952, 5). Of these two areas, extensive evidence for large scale iron production has been discovered only in the Northants region.

Although only one non-slag tapping furnace dated to between 700 and 500 BC was discovered close to the settlement of Corby 2 (Knight, 1984, 165-167), the slag discovered at Castle Yard, and the material remains recovered from Hunsbury would suggest a considerable emphasis on iron production in the Northants region. The preliminary analysis, by Dr. G. MacDonnell of a quantity of slag discovered at Castle Yard reveals that the slag was a product of a slag-tapping furnace (Knight, 1984, 166 & 188). This slag has been dated to as early as 450 BC and, if correct, would indicate that the iron smelting technology of the Northants region was more advanced than the rest of Britain insofar as present evidence goes.

The material remains recovered from the Hunsbury Hillfort reveal both an extensive production of iron artefacts on the site, and the very high status of the resident Iron Age society. Knight believes that Hunsbury was a centralized, non-subsistence production centre, his view being based on the large quantities of bronze, iron, quern, and textile manufacturing articles discovered there (Knight, 1984, 327). The finds of imported lead, jet, shale, and glass in addition to an extensive collection of horse-trappings and weaponry also emphasize Hunsbury's wealth. Knight concludes for post-450 BC Hunsbury that:

> we might envisage a complex series of linkages between Hunsbury and neighbouring settlements, involving perhaps the importation of agricultural products and/or raw materials (e.g. for iron smelting), the export of finished goods (e.g. querns), and (given the evidence for a strong warrior element) possibly even labour services or tribute (Knight, 1984, 329).

Tylecote (1962, 193) states that the quantity of slag discovered at Hunsbury is indicative of smelting, and iron ore could have been easily collected from pits penetrating the ironstone beneath the site or from nearby workings. Additional evidence of Hunsbury's metallurgical industry is the large number of blacksmithing tools, weapons and unfinished goods discovered there.

The tools discovered at Hunsbury cover the complete range of the blacksmith's assemblage described in Chapter 2: two hammers, one file, and four chisels suitable for metalworking. Possible fullers, anvils, and blacksmith's pokers (HNY18b, HNY19a, HNY60a-HNY61b, and HNY65a) were also part of the Hunsbury assemblage. A number of artefacts discovered during the quarrying of the hill could also be classified as either unfinished iron objects, or as objects resulting from mass production. The first artefact of importance is currency bar handle HNY8a. HNY8a reveals all the roughness of an artefact abandoned during forging. The plates of the handle are incompletely welded and eratically sized. The width of the blade varies and has unfinished edges. Moreover, the ends of both the socket and the partial blade do not possess the straight cuts a currency bar usually exhibits when cut for stock. The phosphorus concentration of the bar is also unusually low for its type. In addition, the body of the handle is roughly formed, possibly indicating that the product was manufactured either by a poorly skilled blacksmith, or a blacksmith's apprentice.

Artefacts HNY48b to HNY50b are representative of a group of 50 iron objects. All range from 8 to 15mm in diameter, and 100 to 200mm long, and can be classified either as simple cut-off scraps produced during forging, or as iron rods cut into regular lengths during the production of a chain. Thus, these artefacts could be viewed as the scrap of a large iron industry, or as representing the earlier process in the manufacturing of a metal intensive object.

Artefact HNY52b implies that the mass production of iron tools could have been performed at Hunsbury. The artefact consists of two identical ploughshares rusted together, one inside the socket of the other. This method of stacking artefacts is usually employed by metalworkers for storing their wares, as opposed to farmers saving their supplies. It would seem unlikely that a farmer would possess more than one of these metal intensive objects. If so, these two ploughshares can be seen possibly as having been stored by a blacksmith for eventual trade.

The final artefact of interest discovered at Hunsbury is a broken dagger rusted in the lip of a well manufactured and complete sword scabbard chape. These two artefacts were undoubtedly scrap that was being collected by a blacksmith for recycling.

The dating of Hunsbury suggests that it was inhabited principally during the Middle-Late Iron Age. Much of the bronze work discovered at Hunsbury dates from Danebury Ceramic Phase 7 and later, which corresponds with the time of the increase in phosphoritic iron. The site reveals all the indications of a large iron manufacturing centre. Hunsbury's close proximity to rich, high-phosphorus iron ore would also make it a likely candidate for the source of the high-phosphorus iron which was so prevalent by the Middle Iron Age. Thus, Hunsbury could have been a central site developing around an iron industry based on the trade of phosphoritic iron. The only question that still remains is why were most of the artefacts recovered from Hunsbury low in phosphorus. One possible explanation could be that phosphoritic iron was considered to be low-quality stock. The larger, more prestigious sites were able to obtain the low-phosphorus iron, whereas the non-elite settlement sites would be provided with the more generally available phosphoritic iron. Hunsbury, as an elite site, thus, should have predominantly low-phosphorus iron artefacts.

One might also expect that blacksmithing technology was affected during the 1st century BC by Roman encroachment on the Continent and trade instigated at a distance by Roman entrepreneurs. It was during this period that smelting technology was advanced in south-eastern Britain by the introduction of the slag-tapping B.1 furnaces (Cleere, 1981, 185-186). The only data available indicating that blacksmithing technology may also have been affected are the sudden specialized use of phosphoritic iron evident in both the Bigbury assemblage, and the apparent decrease in the quantity of phosphoritic iron employed in Ceramic Phase 9, which were the result of the proper selection of an iron alloy for the artefact manufactured, and not to the exploitation of a particular iron stock just because of generally availability. Insufficient artefacts have been sampled from this period and area, however, to indicate if this was the result of external influences.

6.4 Summary

The data accumulated for this thesis have permitted the following conclusions concerning the development and diffusion of British Iron Age iron technology to be formulated. During the period of Ceramic Phases 1-5, the distribution of early furnaces and high-cobalt, high-nickel iron would indicate that iron smelting was being performed predominantly in the Southwestern region of Britain. Blacksmithing was performed throughout Southern Britain, but the technology was limited. Such artefacts as the Llyn Fawr sickle reveal the skill possessed by Iron Age iron-workers in the forming of artefacts, however, techniques for improving a tool's properties were unknown.

The increase in phosphoritic iron during Ceramic Phase 6 would indicate that large scale iron smelting had begun in the Northamptonshire region. From Ceramic Phase 6 to Ceramic Phase 7, high-phosphorus iron was a generally available stock exploited predominantly by the smaller settlement sites. The low-phosphorus iron, though, was being traded into the hillforts, and redistributed from there. There are indications (i.e. the chisels) that a small segment of the blacksmithing community had discovered the technique of either carburization or bloom selection. The process by which carbon was added to iron, however, was considered propriatory, and not distributed generally.

Finally, iron was being smelted at all of the major ore sources of Southern Britain in Ceramic Phases 8 and 9. Although the use of carburization or bloom selection was still restricted, blacksmiths may have discovered that phosphoritic iron possessed an increased hardness and were exploiting it for those tools requiring enhanced properties.

CHAPTER 7: RETROSPECT AND PROSPECT

The objective of this publication was to resolve some of the problems and inconsistencies posed by the previous studies. The compilation of the metallurgical database has successfully increased our understanding of many aspects of Iron Age iron technology. First, the possibility of the preferential use of high-carbon and phosphoritic iron alloys for specific tools could be addressed. Secondly, the question of whether prehistoric blacksmiths actually understood the proper applications of such hardening techniques as cold-working and quenching could be ascertained. Thirdly, the evolution of the sources of iron production, especially those of high-phosphorus, and cobalt-nickel iron, could be traced. Finally, a simple model of the organization of the Wessex ironworking community could be devised.

Although many of the problems and inconsistencies produced by the previous studies of the state of Iron Age iron technology could be resolved by this work, new questions have also been posed. Future research should be directed toward comparing the data so far compiled for the Wessex region with the rest of southern Britain. The source of the high-cobalt, high-nickel iron deserves additional attention. Artefacts from the Devon, Cornwall, and Somerset regions should be sampled to determine whether this type of iron was more prevalent in the south-west, and if it was exploited consistently by this region throughout the Iron Age, or decreased in use with the increase of phosphoritic iron.

Sites surrounding Hunsbury in Northamptonshire should also be sampled to determine whether a similar division in the use of low-phosphorus iron for elite sites and phosphoritic iron for smaller settlement sites occurred in this area as well.

The data collected from the Bigbury assemblage would recommend that other sites from this time period and area should also be investigated to determine if the preferential selection of phosphoritic iron was common and the result of external influences. It is unfortunate, however, that few well-stratified, large iron assemblages dating from the Late Iron Age have been discovered.

Finally, the extension of the study of iron technology from the Iron Age to the period of the Roman Invasion would be enlightening. A comparison of the sophistication of iron technology in Romano-British Wessex with the results of this study could answer the question of whether new techniques for enhancing the properties of iron were generally introduced, or if the only advancement was the expansion of the iron industry.

Now that an orderly and comprehensive database of Iron Age iron technology is available, new models can be developed and tested to further increase our understanding of the method by which ironworking expanded and developed during the late prehistoric period in Britain.

APPENDIX A: THE DESCRIPTIONS OF THE SITES SELECTED

Appendix A offers a summary of the evidence relevant to the sites from which artefacts were sampled. The heading line gives site name, site code, and National Grid Reference. The body of the entry outlines the site type, its location, period of occupation if known, a summary of the ironwork recovered and sampled, and any evidence of smelting or forging which may have been found on the site. The entries conclude with the present location of the artefacts sampled, and the major references. The sites are listed alphabetically by name.

All Cannings Cross, All Cannings, Wilts (ACC) SU 080634

An Early Iron Age site situated approximately six miles east of Devizes. No enclosure ditches or other delineations were discovered during excavation. The site has been dated on the basis of the pottery collected to the Early Iron Age, or Danebury Ceramic Phases 1-2.

A total of thirty iron artefacts were discovered at All Cannings Cross. Nine were sampled: one chisel, one small hooked cutting blade, two awl shaped rods, and five blades. These artefacts were selected on the basis of availability, and robustness.

A small quantity of Lower Greensand iron ore, and iron slag were also discovered. The actual weights were not listed in the report. It is doubtful from the description whether any smelting would have occurred.

The artefacts and the samples are presently situated in the Devizes Museum.

References:

Cunnington, 1923a.

Barbury Castle, Wroughton, Wilts (BC) SU 149763

An oval-shaped, bivallate hillfort encompassing an area of 11.6 hectares. The site has not been systematically excavated. Barbury Castle probably dates to Ceramic Phases 4-7.

No records concerning the discovery of the artefacts exist. All that is known is that 26 iron artefacts were donated by the Rev. Henry Harris to the Marlborough College Museum in 1875.

Thirteen artefacts were sampled: 1 blade, 5 hooked blades, 3 spearheads, 3 awl shaped rods, and a large pyramidal iron ingot or anvil. The presence of a hammer (now lost) similar to the Bredon example, and the large ingot or anvil would imply that forging was performed on the site. Since the site was not excavated, the extent of metalworking and the possibility of smelting cannot be ascertained.

The artefacts and the samples are now in Devizes Museum.

References:

MacGregor & Simpson, 1963.

Battery Hill Street, Winchester, Hants (BTH) SU 465265

In 1927, construction of Council housing in Winchester uncovered an inhumation burial along what is now Battery Hill Street. A Mr. Sydney Ward-Evans, who was a local antiquary, collected the resulting artefacts. The objects consist of human bones, a black ware urn, a knife, and a small hooked blade. Both the knife and the hooked blade were sampled.

The artefacts are presently situated in the Winchester City Museum. The samples are in the British Museum.

References:

Unpublished.

Battlesbury Camp, Warminster, Wilts (BTC) ST 897460

A multivallate hillfort encompassing an area of 9.7 hectares. Nine pits were uncovered and excavated when a cistern was installed in 1922. The artefacts are believed to date from Ceramic Phase 6-7, though predominantly 7.

Nine iron artefacts were recovered. Six objects were sampled: four blades, a saw, and a small hooked blade. No evidence of metalworking or smelting was recovered from the site.

The artefacts and the samples are now in the possession of the Devizes Museum.

References:

Cunnington, 1923b.

Bigbury, Harbledown, Kent (B) TR 117575

A univallate hillfort encompassing an area of 26.4 hectares. The site is situated approximately 2 miles west of Canterbury. A series of trenches were excavated in three areas of the hillfort: through the rampart, along the crest of the rampart, and in the annex. The occupation of the site is equivalent to Ceramic Phase 9.

Approximately 60 iron artefacts were recovered during 19th century gravel-quarrying. Twenty-three iron artefacts were sampled: nine hooked cutting blades, one adze, one pick-axe, one axe, two ploughshares, one chisel, one blade, one spearhead, one dagger, one cauldron hook, one bridle bit, one bronze covered ring, one slave chain, and one hammer.

A small quantity of slag was discovered on the site. It was reported by Jessup and Cook (1936) to be forging slag.

The artefacts sampled and the samples are presently situated in the Manchester University Museum. Additional unsampled artefacts are in the Canterbury Museum, and the Maidstone Museum.

References:

Jessup & Cook, 1936. Thompson, 1984.

Buckland Rings, Hants (BR) SZ 314968

A multivallate hillfort encompassing an area of 7.9 hectares. The site is situated approximately one mile north of Lymington. A series of trenches were excavated across the ramparts, and the entrance way. The pottery types discovered at Buckland Rings would date it to the later pre-Roman Iron Age (equivalent to Ceramic Phases 7-8).

Five iron artefacts were discovered during the excavation. Only two were sampled: a possible blade fragment and a sheet. There is no mention in the site report of any slag or iron ore being discovered on the site.

The artefacts are in Winchester City Museum. The samples are in the British Museum.

References:

Hawkes, 1936.

Bury Hill, Blatford, Hants (BH) SU 346435

A two phase multivallate hillfort encompassing an area of 54.3 hectares. The site is situated 1.5 miles southwest of Andover. Bury Hill was excavated in conjunction with Balksbury, which is approximately a half-mile northeast of Bury Hill. A series of excavations were carried out on the ramparts of both Camps, and the entrance way. Three phases were discovered: Bury Hill I (equivalent to Ceramic Phase 1-3); Bury Hill II (equivalent to Ceramic Phases 6 and 7); and Bury Hill III (equivalent to Ceramic Phase 8-9).

Four iron artefacts were discovered at Bury Hill. All were sampled: a scabbard attachment from Bury Hill II; and a bar, a ring, and a nail from Bury Hill III. There was no mention of either slag or iron ore being discovered on the site.

The artefacts are in Winchester City Museum. The samples are in the British Museum.

References:

Hawkes, 1940.

Casterley Camp, Upavon, Wilts (CC) SU 115535

An univallate hillfort encompassing an area of 61.8 hectares. The site is situated approximately two miles southwest of Upavon. Sections of the rampart, the entrances, and the internal rectangular enclosure were excavated. The pottery discovered on the site would indicate an occupation dating from the equivalent of Ceramic Phases 8-9.

Thirty-eight iron artefacts were discovered. Only two were sampled: an axe and a small blade. There is no mention of slag or other forging or smelting evidence.

The artefacts and the samples are in Devizes Museum.

References:

Cunnington & Cunnington, 1913.

Cow Down, Longbridge, Deverell (CD) ST 887145

An Early Iron Age site situated approximately 2 miles south of Warminster. A series of three ditched enclosures were excavated. Enclosure B, from which the blade originated, encompassed an area of approximately 17 hectares. The material recovered from the excavation would indicate that the site was occupied during the Early Iron Age, or Ceramic Phases 1-3.

Only the one hooked blade sampled was discovered during excavation. It was found in a pit on top of a quantity of carbonized grain, and fragments of a sack in which the grain appeared to have been contained.

No evidence of forging or smelting was recovered during excavation.

The hooked blade and the sample are in Devizes Museum.

References:

Annable, 1958, 9-10. Annable, 1961, 31-32.

Danebury, Nether Wallop, Hants (D) SU 323376

A multivallate hillfort encompassing an area of 5.2 hectares. The site is situated 5 km. northwest of Stockbridge, and has been extensively excavated since 1969. The Ceramic Phases determined for Danebury are the standard used in this dissertation:

Ceramic Phases	1-3	550-450 BC
Ceramic Phases	4-5	450-400 BC
Ceramic Phase	6	400-300 BC
Ceramic Phase	7	300-100 BC
Ceramic Phase	8	100 BC-10 AD

Three hundred and six iron artefacts were discovered at Danebury. One hundred and forty-one were sampled. The artefacts sampled include a range of tools, personal items (e.g. finger rings), and scrap. Quantities of slag and iron ore were found at Danebury, but analysis shows that it resulted from forging; no conclusive evidence for smelting was discovered.

The artefacts are presently situated in the Institute of Archaeology at Oxford University. The samples are in the British Museum.

References:

Cunliffe, 1984a.

Fifield Bavant Down, Ebbesbourne, Wilts (FBD) SU 009257

Mr. R.C.C. Clay excavated a series of 107 pits at Fifield Bavant Down in 1922. The pits were revealed on the surface by differential growth of vegetation. The associated pottery would date the site to Ceramic Phases 4 to 6.

Fourteen iron objects were discovered. Two were sampled: a blade and a small hooked cutting blade.

It is difficult to determine from Clay's publication if any forging remains were discovered. Slag is mentioned frequently and also there is a reference to 'a clay hearth...containing...a fused mass of iron, charcoal, and earth' (Clay, 1924, 471). However, no more details are mentioned.

The artefacts and the samples are in Devizes Museum.

References:

Clay, 1924.

Hunsbury, Hardingstone, Northants (H) SP 738583

A univallate hillfort encompassing an area of 1.7 hectares. The site is situated approximately three miles southeast of Northampton. The Hunsbury iron assemblage was recovered during 19th century quarrying for iron ore. No contexts are known for the artefacts.

One hundred and thirty-nine of the approximately 150 iron artefacts from Hunsbury were sampled. Although the analysis of these artefacts was used in the evaluation of both the sophistication of iron technology and the use of phosphoritic iron, the data could not be used in diachronic analysis.

The artefacts are situated in Northampton's Central Museum. The samples are in the British Museum.

References:

Dryden, 1885.
Fell, 1937.
Fell, 1953.
George, 1917.

Isle of Portland (PTLD) SY 691706

The two iron ingots sampled were stray finds. A local antiquary identified the artefacts during a visit to a neighbour's. The ingots were being employed as doorstops. The shapes are sufficiently distinctive, however, that identification of these two iron ingots as the 'spitzbarren' type employed on the Continent during the Iron Age was possible.

The artefacts are in Portland Museum. The samples are in the British Museum.

References:

Tylecote, 1962, 210.

Little Somborne, Hants (LS) SU 389326

An oval-shaped ditched enclosure surrounded by 'celtic' fields. The site is situated on the gentle southern slope of Windmill Hill, and encompasses an area of approximately 7.4 hectares. The excavation was a 140m by 20m section bisecting the site. The features are dated from the fifth to second centuries BC (Ceramic Phases 4-7).

Only three iron artefacts were recovered: two fragments and a finely produced spearhead. The large fragment and the spearhead were sampled.

Small quantities of smithing slag were recovered from two pits and a v-shaped ditch located outside the perimeter of the enclosure. The scarcity of slag would indicate that only light forging was performed on the site. No indications of smelting were found.

The artefacts are in the possession of the Hampshire County Museum Service. The samples are in the British Museum.

References:

Neal, 1979.

Little Waltham, Chelmsford, Essex (LW) TL 705126

Little Waltham is situated approximately 4 miles north of Chelmsford. The site went through three phases of occupation during the Iron Age. The first two phases were the open settlements of Little Waltham II (equivalent to early Ceramic Phase 7), and Little Waltham III (equivalent to Ceramic Phases 7-8). The final occupation phase of Little Waltham was the palisade-enclosed settlement of Little Waltham IV (equivalent to late Ceramic Phase 8).

Thirty iron artefacts were recovered, of which twelve were sampled: two strips, two plates, one bar, one sword scabbard fragment, one rod, one ring, two nail shanks, one bucket attachment, and one staple.

No slag was found, but two fragments of bloomery iron were discovered associated with period III. Professor Tylecote sampled these artefacts and listed them as high phosphorus, low carbon iron, and only lightly forged since smelting.

The artefacts are housed by the Chelmsford Archaeological Trust. The samples are in the British Museum.

References:

Drury, 1978a. Drury, 1978b.

Meon Hill, Warwickshire (MH) SP 177453

A large multivallate hillfort encompassing an area of 10.4 hectares, and lying six miles south of Stratford-upon-Avon. The site has not been systematically excavated.

The five currency bars were part of a hoard discovered and donated to the Warwick Museum in the late 19th century. The exact location in the hillfort from which the bars were found is unknown. All five currency bars were sampled.

The artefacts are in the Warwick Museum. The samples are in the British Museum.

References:

Hodges, 1906.

Nadbury Hillfort, Ratley, Warwickshire (MH3b) SP 390482

Nadbury is a univallate Hillfort encompassing an area of 6.5 hectares. The site is situated approximately 7 miles north of Banbury, on the Edgehill Scarp. A small excavation was completed in 1984 on the outer rampart in advance of road construction. No previous excavation had been performed.

A single currency bar and a nail were found, of which the currency bar was sampled. No evidence of smelting or forging was discovered.

The currency bar is in the Warwick Museum. The sample is in the British Museum.

References:

Unpublished.

Old Down Farm, Andover, Hants (ODF) SU 356465

Old Down Farm is a subrectangular ditched enclosure situated above a tributary of the River Anton on a spur of Upper Chalk. The site has only one entrance and encompasses an area of 1.18 hectares.

The site contains features from the Neolithic, Roman, and Saxon periods, in addition to the extensive Iron Age remains. Smithing residues were discovered in two pits dating from the 3rd to 1st centuries BC. No indications of smelting were found. A total of 26 iron artefacts were sampled: four rods, one wire, eight fragments, two large hooked cutting blades, one small hooked cutting blade, one chisel, two currency bars, one nail, three blades, a ploughshare point, one arrowhead, and one bronze covered circular object.

The artefacts are housed by the Winchester County Museum Service. The samples are in the British Museum.

References:

Davies, 1979. Davies, 1980.

Rushall Down, Upavon, Hants (ACC5b) SU 087521

The Rushall Down has produced an extensive collection of artefacts during ploughing through the centuries. The site was excavated in the 1890's by Lt. Col. W. Hawley. The results and finds of the excavation, however, have never been completely reported. Predominantly, the artefacts appear to date from the Romano-British period, though some 'coarse hand-made pottery' was also discovered.

The hooked blade sampled was one of an unidentified number of iron artefacts. No smelting or forging remains have been uncovered.

The artefact and the sample is in Devizes Museum.

References:

Hawley, 1924. Pugh & Crittall, 1957, 99-100.
Cunnington, 1930, 201.

West Down, Chilbolten, Hants (C) SU 418312

The site is a circular, ditched enclosure encompassing an area of between two and six hectares. The site was not properly excavated, but a Mr. Tucker dug up parts of it in 1919. All of the artefacts recovered were donated to Dr. Williams-Freeman, eventually reaching the Winchester City Museum.

One sword fragment, two blades, and a pair of small shears were sampled from the site. The actual quantity of ironwork discovered is unknown, as is the possibility of smelting or forging having occurred on the site.

The artefacts are in the possession of the Winchester City Museum. The samples are in the British Museum.

References:

Unpublished.

Wilsford Down, North Wilts. (BC7b) SU 104573

The settlement site on Wilsford Down has never been excavated. The site was only identified through pottery uncovered by moles and by surface irregularities. Periodically, artefacts are found accidentally, which are donated to the Devizes Museum. On the basis of pottery discovered, the site was occupied during the Romano-British period, or Ceramic Phase 9.

The hooked blade is one of an unidentified number of iron artefacts, including spearheads, knives, and arrowheads found at Wilsford Down. No indication of smelting or forging have been discovered on the Down.

The artefact and the sample is in Devizes Museum.

References:

Cunnington, 1930, 208. Pugh & Crittall, 1957, 122.

Winnal Down, Winchester, Hants (WND) SU 501309

A D-shaped ditched settlement enclosure encompassing an area of one half hectare. The site provided artefacts from the Neolithic period through to Medieval times. The site is located in eastern Winchester on the river Itchen. A detailed report on the finds of Winnal Down has not yet been published.

The iron artefacts of Winnal Down consist predominantly of nails. Twenty-two nails, two hooked blades, one knife, and three unidentifiable objects were sampled. No indications of iron smelting or forging were found on the site.

The artefacts are being housed by the Winchester County Museum Service. The samples are in the British Museum.

References:

Fasham, 1978.

Worthy Down, Headbourne Worthy, Hants (WD) SU 469350

A settlement site located 3 miles north-west of Winchester. Excavation revealed a series of pits and ditches produced by three phases of inhabitation. Phase One was a casual occupation dating from approximately 500 BC. More extensive occupation occurred, however, during the periods of 150 BC to 100 BC (equivalent to Ceramic Phase 7), and 50 BC to 40 AD (equivalent to Ceramic Phase 8).

Approximately 20 iron artefacts were discovered including seven whole currency bars and fragments of six others. One ploughshare, one wedge, two awls, one blade fragment, one linch-pin, one bar and all of the currency bars were sampled.

Slag and part of a tuyere were discovered in a later pit. The quantity of slag found, however, is unknown, making the manner of production difficult to identify. No indication of smelting or forging is evident from the earlier phases. Forging evidently was practiced on or near the site, indicated by the large number of currency bars, some incomplete.

The artefacts are housed by the Winchester County Museum Service. The samples are in the British Museum.

References:

Hooley, 1921. Hooley, 1931.

APPENDIX B: THE DESCRIPTIONS OF THE ARTEFACTS SAMPLED

This Appendix gives details of the artefacts sampled. Each entry consists of six sections. The first three sections are contained in the heading line, and present, from left to right, the sample's Mount Number, the artefact's Small Finds or Museum Accession Number, and the equivalent Ceramic Phase from which the artefact dates if it can be ascertained. The sample's Mount Number is a letter-number code which specifies both the artefact's site, and the mount in which the sample is contained. The first series of capitalized letters identifies the site from which the artefact came, and are summarized below:

All Cannings Cross	ACC	Isle of Portland	PTLD
Barbury Castle	BC	Little Somborne	LS
Battery Hill	BTH	Little Waltham	LW
Battlesbury Camp	BTC	Meon Hill	MH
Bigbury	B	Nadbury	MH3b
Buckland Rings	BR	Old Down Farm	ODF
Bury Hill	BH	Rushall Down	ACC5b
Casterly Camp	CC	West Down	C
Cow Down	CD	Wilsford Down	BC7b
Danebury	D	Winnal Down	WND
Fifield Bavant Down	FBD	Worthy Down	WD
Hunsbury	H		

The number and letter following the capitalized letters specifies which mount the sample is presently in.

The main body of each entry presents a description of the artefact sampled, including the location from which the sample was taken. The final section of each entry is the artefact's site context and a reference to a publication in which the artefact appears. The Hunsbury publications only presented general photographs of the artefacts with no identification. Thus, references for these artefacts were impossible to trace.

The entries are alphabetically ordered according to Mount Number.

ACC1a cp. 1-3

A socketed chisel measuring 147mm long. The sample was taken from the blade.

Pit G6, 1.5ft deep. Cunnington, 1923a, Plate 20-2.

ACC1b cp. 1-3

A small, tanged, hooked, cutting blade measuring 67.5mm long, and 22.5mm wide at the base of the blade. Two rivets pierce the tang. The only indication of a socket is a slight curvature on one side of the tang. The sample was taken from the cutting edge at the hook.

Pit G8, 1.5ft deep. Cunnington, 1923a, Plate 20-3.

ACC2a Small Finds 790105 cp. 1-3

An awl type rod measuring 87mm long, and 3.7x3.7mm in dimensions at one end tapering to a point. The sample was taken from the pointed end.

Pit F6, 1.25ft deep. Cunnington, 1923a, Plate 20-8.

ACC2b Small Finds 790110 cp. 1-3

A rod measuring 64.5mm long, and 3x4.5mm in dimensions. Both ends are pointed. The sample was taken from the sharper end.

Pit G4. Cunnington, 1923a, Plate 20-6.

ACC3a cp. 1-3

A tanged blade measuring 88.5mm long, and 18mm wide at the base of the blade tapering to 4.5mm wide at the fracture. The artefact is missing a portion of the point. The sample was taken from the cutting edge near the tang.

Pit I3, 1.3ft deep. Cunnington, 1923a, Plate 20-14.

ACC3b cp. 1-3

A blade point measuring 45mm long, and 19.5mm wide at the fracture tapering to a rounded point. The sample was taken from the cutting edge near the fracture.

Pit G5, 1.25ft. deep. Cunnington, 1923a, Plate 20-16).

ACC4a cp. 1-3

A curved tanged blade measuring 84mm long, and 13.5mm wide at the base of the blade tapering to a rounded point. The sample was taken from the cutting edge near the tang.

Pit K4, 2ft deep. Cunnington, 1923a, Plate 20-13.

ACC4b cp. 1-3

A tanged, blade measuring 69mm long, and 13.5mm wide at the base of the blade tapering to a rounded point. The sample was taken from the cutting edge at the hook.

Pit K4, 2ft deep. Cunnington, 1923a, Plate 20-11.

ACC5a cp. 1-3

A blade point measuring 61.5mm long, and 28.5mm wide at the fracture tapering to a rounded point. The sample was taken from the cutting edge near the fracture.

Pit H8, 2ft deep. Cunnington, 1923a, Plate 20-12.

ACC5RD cp. -

A large, hooked, cutting blade measuring 130mm long, and 35mm wide at the fracture. The sample was taken from the cutting edge between the point and the hook.

Pugh & Crittall, 1957, 100.

B1a Museum Acc. 35831 cp. 9

A hooked blade fragment measuring 168mm long, and 28mm wide at the fracture. The socket and the point of the artefact are missing. The sample was taken from the cutting edge at the hook.

Thompson, 1984, Fig. 14-12.

B1b Museum Acc. 35820 cp. 9

A socketed, hooked blade measuring 204mm long, and 52mm wide at the base of the blade. A rivet hole pierces the socket between the flanges. The sample was taken from the cutting edge at the base.

Thompson, 1984, Fig. 14-9.

B2a Museum Acc. 35819 cp. 9

A socketed, hooked blade measuring 208mm long, and 56mm wide at the base of the blade. A rivet hole pierces the socket between the flanges. The sample was taken from the cutting edge at the base of the blade.

Thompson, 1984, Fig. 14-10.

B2b Museum Acc. 35814 cp. 9

A socketed, hooked blade measuring 260mm long, and 56mm wide at the base of the blade. A rivet hole pierces the socket between the flanges. The sample was taken from the cutting edge at the hook.

Thompson, 1984, Fig. 14-11.

B3a Museum Acc. 35823 cp. 9

A tanged, hooked blade measuring 432mm long, and 44mm wide at the base of the blade. The tang measures 246mm long. The sample was taken from the cutting edge at the hook.

Thompson, 1984, Fig. 19-51.

B3b Museum Acc. 35809 cp. 9

A hammer measuring 120mm long, 24mm wide just in front of the socket flaring to 28mm wide at the head, and 24mm thick at the socket tapering to 16mm thick at the head. The sample was taken from the head.

Thompson, 1984, Fig. 14-17.

B4a Museum Acc. 35813 cp. 9

A pick-axe measuring 280mm long, 24mm wide just in front of the socket tapering to an 8mm wide vertical blade, 40mm wide behind the socket flaring to a 48mm wide horizontal blade, and 36mm thick in front of the socket flaring to 48mm wide at the vertical blade, and 24mm thick behind the socket tapering to a 16mm thick horizontal blade. The sample was taken from the vertical blade.

Thompson, 1984, Fig. 14-16.

B4b Museum Acc. 35813 cp. 9

A second sample was taken from the artefact described for B4a. The sample was taken from the horizontal blade.

Thompson, 1984, Fig. 14-16.

B5a Museum Acc. 35812 cp. 9

An axe head measuring 220mm long, 32mm wide in front of the socket tapering to a sharp blade, and 52mm thick in front of the socket flaring to 64mm at the blade. The sample was taken from the blade.

Thompson, 1984, Fig. 14-15.

B5b Museum Acc. 35824 cp. 9

A ploughshare measuring 316mm long, 52mm wide at the base of the blade tapering to a blunt point, and 16mm thick at the base of the blade tapering to a blunt point. The sample was taken from the point.

Thompson, 1984, Fig. 19-50.

B6a Museum Acc. 35825 cp. 9

A ploughshare measuring 356mm long, 40mm wide at the base of the blade tapering to a sharp point, and 16mm thick at the base of the blade tapering to a sharp point. The sample was taken from the point.

Thompson, 1984, Fig. 19-49.

B6b Museum Acc. 35811 cp. 9

A chisel measuring 232mm long, 28mm wide at the solid butt tapering to 16mm at the blade, and 28mm thick at the butt tapering to a sharp blade. The butt possesses a burr from being struck with a hammer. The sample was taken from the blade.

Thompson, 1984, Fig. 19-52.

B7a Museum Acc. 35818 cp. 9

A hooked blade fragment measuring 292mm long, and 68mm wide at the fracture. The socket or tang of the blade is missing. The sample was taken from the cutting edge at the hook.

Thompson, 1984, Fig. 14-14.

B7b Museum Acc. 35822 cp. 9

A socketed, hooked blade measuring 384mm long, and 80mm wide at the base of the blade. Most of the socket and the point of the artefact are missing. The sample was taken from the cutting edge at the base of the blade.

Unpublished.

B8a Museum Acc. 35815 cp. 9

A socketed, hooked blade measuring 368mm long, and 56mm wide at the base of the blade. A rivet pierces the socket between the flanges. The sample was taken from the cutting edge at the base of the blade.

Unpublished.

B8b Museum Acc. 35816 cp. 9

A socketed, hooked blade measuring 348mm long, and 68mm wide at the base of the blade. A rivet pierces the socket between the flanges. The sample was taken from the cutting edge at the base of the blade.

Thompson, 1984, Plate I-3.

B9a Museum Acc. 35817 cp. 9

A socketed, hooked blade measuring 336mm long, and 52mm wide at the base of the blade. A rivet hole pierces the socket between the flanges. The sample was taken from the cutting edge at the base of the blade.

Unpublished.

B9b Museum Acc. 35808 cp. 9

A tanged blade measuring 141mm long, and 14mm wide at the base of the blade tapering to the point. The sample was taken from the cutting edge at the base of the blade.

Unpublished.

B10a Museum Acc. 35805 cp. 9

A dagger measuring 292mm long, and 24mm wide at the base of the blade tapering to the fracture. The point of the dagger is missing. The sample was taken from one of the cutting edges.

Thompson, 1984, Fig. 19-53.

B10b Museum Acc. 35804 cp. 9

A socketed javelin head measuring 91mm long, and 18mm wide at the base of the blade. The point and most of the edges of the blades are missing. The sample was taken from one edge near the socket.

Unpublished.

B11a Museum Acc. 35789 cp. 9

A cauldron hook consisting of a ring measuring 108mm in diameter attached to a 292mm long twisted iron shaft, attached to four 120mm long, 32mm wide chain links. The sample was taken from the shaft.

Thompson, 1984, Fig. 16-26.

B11b Museum Acc. 35800 cp. 9

A bridle-bit fragment consisting of three, attached, corroded 80mm long, 18mm wide links. The sample was taken from the end of one of the links.

Thompson, 1984, Fig. 17-35.

B12a Museum Acc. 35801 cp. 9

A bronze covered harness ring measuring 94mm in diameter, and produced from a piece of metal 11mm in diameter. A portion of the radius is missing. The sample was taken from an intact, exterior edge.

Unpublished.

B12b Museum Acc. 35795 cp. 9

A slave-chain measuring approximately 2000cm long and consisting of six neck braces attached by chain-links. The sample was taken from one link.

Thompson, 1984, Fig. 18-45.

B13a Museum Acc. 35810 cp. 9

A hammer measuring 84mm long, 16mm wide, and 16mm thick at the butt tapering to 8mm at the head. The butt of the artefact has a burr from being struck, which might indicate that the tool was actually a hot-chisel. The sample was taken from the blade.

Thompson, 1984, Fig. 14-18.

B13b Museum Acc. 35810 cp. 9

A second sample was taken from the artefact described for B13a. The sample was taken from the butt end of the artefact.

Thompson, 1984, Fig. 14-18.

BC1a Catalogue No. 1 cp. 4-7

A blade measuring 292mm long, and 40mm wide at the handle tapering to a rounded point. The sample was taken from the cutting edge.

MacGregor and Simpson, 1963, Fig. 1-1.

BC1b Catalogue No. 2 cp. 4-7

A large, socketed, hooked cutting blade measuring 146mm long, and 38mm wide at the base of the blade. The sample was taken from the cutting edge between the point and the hook.

MacGregor and Simpson, 1963, Fig. 1-2.

BC2a Catalogue No. 3 cp. 4-7

A large, tanged, hooked cutting blade measuring 138mm long, and 30mm wide at the base of the blade. The sample was taken from the cutting edge at the hook.

MacGregor and Simpson, 1963, Fig. 1-3.

BC2b Catalogue No. 4 cp. 4-7

A large, socketed, tanged, hooked cutting blade measuring 136mm long, and 30mm wide at the base of the blade. The sample was taken from the cutting edge between the point and the hook.

MacGregor and Simpson, 1963, Fig. 1-4.

BC3a Catalogue No. 5 cp. 4-7

A small, socketed, hooked cutting blade measuring 68mm long, and 18mm wide at the base of the blade. The sample was taken from the cutting edge at the hook.

MacGregor and Simpson, 1963, Fig. 1-5.

BC3b Catalogue No. 6 cp. 4-7

A small, socketed, hooked blade measuring 65mm long, and 20mm wide at the base of the blade. The sample was taken from the cutting edge at the hook.

MacGregor and Simpson, 1963, Fig. 1-6.

BC4a Catalogue No. 8 cp. 4-7

A spearhead fragment measuring 92mm long, 6mm wide at the fracture at the base of the blade, and 24mm wide at the centre of the blade tapering to a rounded point. The artefact measures 3mm thick along its central axis. The socket of the spearhead is missing. The sample was taken from one edge near the point.

MacGregor and Simpson, 1963, Fig. 1-8.

BC4b Catalogue No. 10 cp. 4-7

A socketed, midribbed, leaf-shaped spearhead measuring 108mm long, 10mm wide at the top of the socket, and 26mm wide at the centre of the blade tapering to a rounded point. A single rivet bisects the base of the socket. The sample was taken from the widest section of the blade at the edge.

MacGregor and Simpson, 1963, Fig. 1-10.

BC5a Catalogue No. 11 cp. 4-7

A socketed, leaf-shaped spearhead measuring 138mm long, 10mm wide at the top of the socket, and 30mm wide at the centre of the blade tapering to a rounded point. The artefact measures 4mm thick at the mid-rib tapering to still sharp edges on both sides. The spearhead is missing a large portion of the socket. The sample was taken from the cutting edge at the widest section of the blade.

MacGregor and Simpson, 1963, Fig. 1-11.

BC5b Catalogue No. 19 cp. 4-7

A rod measuring 110mm long, and 4x8mm in dimensions. The sample was taken from one end.

MacGregor and Simpson, 1963, Fig. 2-19.

BC6a Catalogue No. 21 cp. 4-7

A curved rod fragment measuring 56mm long, and 4mm in diameter at one end flattening out to a 6x2mm cross-section at the other. The sample was taken from the flattened end.

MacGregor and Simpson, 1963, Fig. 2-21.

BC6b Catalogue No. 22 cp. 4-7

An awl shaped rod measuring 56mm long, 7mm wide, and 2mm thick. One end of the artefact is pointed. The sample was taken from the pointed end.

MacGregor and Simpson, 1963, Fig. 2-22.

BC7a Catalogue No. 26 cp. 4-7

An anvil fragment measuring 130mm long, and 72x58mm in dimensions at the base tapering to a curved point. The sample was taken from the bottom edge of the object.

MacGregor and Simpson, 1963, Fig. 2-26.

BC7WD cp. R-B

A large, tanged, hooked cutting blade measuring 160mm long, and 29mm wide at the base of the blade. The sample was taken from the cutting edge at the hook.

Pugh & Crittall, 1957, 122.

BH1a Small Finds 6.1 cp. -

A rod measuring 51mm long, 12mm wide and 7mm thick. The sample was taken from one end.

Unpublished.

BH1b Small Finds 6.5 cp. -

An object measuring 60mm long, and varing in width from 26mm at the base, to 18mm near the centre, to 30mm in the middle, to 8mm at the opposite end. The wider end is curved, and a rivet hole pierces the centre of the object. The artefact is believed to be a section of a scabbard-loop. The sample was taken from an edge near the wider end.

Ditch II. Hawkes, 1940, Fig. 17.

BH2a Small Finds 6.8 cp. -

A heavy ring produced from a strip measuring 11mm wide, and 4-7mm thick. The ring measures 35mm in diameter. The sample was taken from one edge.

Unpublished.

BH2b Small Finds 6.9 cp. -

A nail measuring 53mm long with a forged flattened head. The sample was taken from the head.

Unpublished.

BR1a Small Finds 5.1 cp. -

A blade fragment measuring 40mm long, and 38mm wide. The sample was taken from the cutting edge.

Occupation-hollow. Hawkes, 1936, 155.

BR1b Small Finds 5.4 cp. -

A roughly rectangular sheet measuring 24-28mm long, 18-20mm wide, and 1.5mm thick. The sample was taken from one edge.

Occupation-hollow. Hawkes, 1936, 155.

BTC1a cp. 7

A saw measuring 208mm long, 28mm wide, and 2mm thick. Two rivets are in the tang. The teeth may have been set to cut on the pull stroke. One tooth was removed for examination.

Pit 9. Cunnington, 1923b, Fig. 7.

BTC1b cp. 7

A small, hooked, cutting blade measuring 70mm long, and 8mm wide at the base of the blade. The sample was taken from the cutting edge between the point and the hook.

Pit 9. Cunnington, 1923b, Fig. 6.

BTC2a cp. 7

A tanged blade measuring 124mm long, and 28mm wide at the base of the blade, slightly tapering to a rounded point. A rivet pierces the tang. The sample was taken from the blade near the tang.

Pit 1. Cunnington, 1923b, Fig. 2.

BTC2b cp. -

A blade fragment measuring 192mm long, and 40mm wide at one end tapering to 22mm at the other. The point and the tang of the blade are missing. The sample was taken from the cutting edge near the wider end.

Pit 8. Cunnington, 1923b, 370.

BTC3a cp. -

A small, roughly rectangular blade fragment measuring 65mm long, and 41mm wide. The sample was taken from the cutting edge near one end.

Pit 8. Cunnington, 1923b, 370.

BTC3b cp. -

A blade fragment measuring 55mm long, and 28mm wide at one end tapering slightly to the other. The point and the tang of the blade are missing. The sample was taken from the cutting edge.

Unpublished.

BTH1a Small Finds 1491.1 cp. -

A tanged blade measuring 123mm long and 28mm wide at the base of the blade tapering to 5mm at the point. The sample was taken from the cutting edge.

Unpublished.

BTH1b Small Finds 1491.2 cp. -

A small, socketed, hooked, cutting blade measuring 73mm long, and 15mm wide at the base of the blade. The sample was taken from the back of the blade.

Unpublished.

Cla Small Finds 303.1 cp. -

A number of sheet fragments of which one was sampled. The roughly rectangular artefact measures 64mm long, 32mm wide, and tapers from 5mm to 1mm thick. The sample was taken from the thick end.

Unpublished.

Clb Small Finds 303.4 cp. -

A narrow blade measuring 90mm long, and 9mm wide. The sample was taken from the cutting edge.

Unpublished.

C2a Small Finds 303.7 cp. -

A single blade from a pair of miniature shears was discovered. The blade measures 29mm long, and 9mm wide at the base of the blade tapering to 3mm at the rounded point. The remains of the shaft measures 44mm long and 3mm in diameter. The sample was taken from the cutting edge.

Unpublished.

C2b Small Finds 303.8 cp. -

A blade point measuring 39mm long, and 14mm wide. The sample was taken from the cutting edge.

Unpublished.

CCla cp. 8-9

An axe blade measuring 86mm long, 54mm wide at the blade tapering to 25mm wide at the fracture, and 9mm thick at the fracture tapering to a still sharp edge. The socket and the butt of the axe are missing. The sample was taken from the cutting edge. The sample disintegrated during polishing.

Ditch 6, 1ft. deep. Cunnington & Cunnington, 1913, Plate III-13.

CC1b cp. 8-9

A small, tanged blade measuring 59mm long, and 17mm wide at the base of the blade tapering to a rounded point. The sample was taken from the cutting edge near the tang.

Ditch 1, 4ft. deep. Cunnington & Cunnington, 1913, Plate II-6.

CD1a cp. 1-3

A large, socketed, tanged, hooked cutting blade measuring 167mm long, and 34mm wide at the base of the blade. The sample was taken from the cutting edge at the hook.

Annable, 1961, 32.

D1a Small Finds 1300 cp. 7

A socketed spearhead measuring 173mm long, and 14mm wide at the top of the socket flaring to 21.5mm at the widest portion. The blade possesses both a raised midrib and two decorative bronze rivets, one in each blade. One rivet hole also pierces the base of the socket. The sample was taken from the blade just below the bronze rivets.

Layer 456. Cunliffe, 1984a, Fig. 2.100.

D1b Small Finds 1303 cp. 6

A socketed spearhead measuring 110.5mm long, and 16mm wide at the top of the socket flaring to 19mm at the widest portion. The blade possesses a raised midrib. Two sets of rivet holes are on opposite sides of the socket. The sample was taken from the lower portion of the blade just above the socket.

Layer 474. Cunliffe, 1984a, Fig. 2.101.

D2 Small Finds 1390 cp. 7-8

A large, socketed, hooked cutting blade measuring 125mm long, and 38mm wide at the base of the blade. The point of the blade is missing. One rivet and one rivet hole pierce the centre of the socket between the two flanges. The sample was taken from the cutting edge at the hook.

Pit P1078 Layer 10. Cunliffe, 1984a, Fig. 2.1.

D3a Small Finds 1297 cp. -

A socketed spearhead measuring 88mm long, and 14mm wide at the top of the socket flaring to 28mm at the widest portion. Two rivet holes pierce the base of the socket, one on each flange. The sample was taken from the cutting edge.

Unstratified. Cunliffe, 1984a, Fig. 2.103.

D3b Small Finds 852 cp. -

A large, socketed, tanged, hooked cutting blade measuring 169mm long, and 38mm wide at the base of the blade. The sample was taken from the cutting edge near the point.

Layer 255. Cunliffe, 1984a, Fig. 2.12.

D4a Small Finds 1247 cp. 3

A small blade measuring 70mm long, and 20mm wide at the base of the blade tapering to 8mm at the fracture. The end of the tang and the point of the blade are missing. The sample was taken from the cutting edge near the tang.

Pit P939 Layer 2. Cunliffe, 1984a, Fig. 2.32.

D4b Small Finds 542 cp. 6

A blade measuring 143mm long, and 25mm wide at the base of the blade tapering to 10mm at the point. The sample was taken from the cutting edge near the handle.

Layer 4. Cunliffe, 1984a, Fig. 2.25.

D5a Small Finds 514 cp. -

A saw blade measuring 133mm long, and 22mm wide at the fracture tapering to 5mm at the tip. A large portion of the teeth are set to cut on the pull stroke. One of the teeth was removed for examination.

Pit P465 Layer 2. Cunliffe, 1984a, Fig. 2.43.

D5b Small Finds 865-866 cp. 7

A large, hooked cutting blade measuring 122mm long, and 30mm wide at the base of the blade. The artefact is broken into two pieces and is missing a portion of the point. One rivet hole pierces the socket between the flanges. The sample was taken from the cutting edge at the hook.

Pit P827 Layer 5. Cunliffe, 1984a, Fig. 2.2.

D6 Small Finds 314 cp. SXN

A Saxon blade measuring 190mm long, and 16mm wide. The sample was taken from the cutting edge.

Pit 252 Layer 1. Cunliffe, 1984a, Fig. 2.26.

D7a Small Finds 259 cp. 7

A blade fragment measuring 177mm long and 20mm wide at the fracture tapering slightly to a still sharp point. The sample was taken from the cutting edge near the fracture.

Pit P110 Layer 7. Cunliffe, 1984a, Fig. 9:F1.

D7b Small Finds 132 cp. 3

A tanged chisel measuring 86mm long, and tapering from 29mm wide at the cutting edge to 3mm at the top of the tang. The sample was taken from the cutting edge.

Pit P70 Layer 5. Cunliffe, 1984a, Fig. 2.47.

D8a Small Finds 358 cp. -

A large, tanged, socketed hooked cutting blade measuring 159mm long and 35mm wide at the base of the blade. The sample was taken from the cutting edge at the hook.

Unstratified. Cunliffe, 1984a, Fig. 2.13.

D8b Small Finds 551 cp. 7

A small, socketed, hooked cutting blade measuring 71mm long, and 20mm wide at the base of the blade. The artefact is broken into two pieces, and is missing both the point and part of the socket. The sample was taken from the cutting edge at the hook.

Layer 5. Cunliffe, 1984a, Fig. 2.11.

D9 Small Finds 771 cp. 3

A socketed spearhead measuring 97mm long, and 22mm wide at the base of the blade tapering to a point. The artefact was crudely fashioned from a broken saw blade. One of the teeth was removed for examination.

Pit P713 Layer 2. Cunliffe, 1984a, Fig. 2.102.

D10 Small Finds 184 cp. 7

A socketed ploughshare measuring 118mm long and tapering from 48mm wide at the base of the socket to 10.5mm at the tip. The sample was taken from the tip.

Pit P94 Layer 3. Cunliffe, 1984a, Fig. 2.71.

D11a Small Finds 485 cp. 7

A socketed chisel measuring 168mm long. The butt of the socket possesses a burr from being struck with a hammer. The sample was taken from the blade.

Pit P452 Layer 1. Cunliffe, 1984a, Fig. 2.44.

D11b Small Finds 612 cp. 7

A socketed chisel measuring 177mm long. The sample was taken from the blade.

Pit P23 Layer 3. Cunliffe, 1984a, Fig. 2.45.

D12 Small Finds 541 cp. 6

A bucket attachment consisting of a wire loop welded to a flattened plate pierced by two rivet holes. The artefact measures 72mm long with a wire cross-section of 7x7mm. The sample was taken from the plate.

Layer 4. Cunliffe, 1984a, Fig. 2.165.

D13a Small Finds 437 cp. 7

A small, socketed, hooked cutting blade measuring 57mm long and 16mm wide at the base of the blade. A single rivet hole pierces the socket between the two flanges. The sample was taken from the cutting edge at the hook.

Pit P337 Layer 10. Cunliffe, 1984a, Fig. 2.8.

D13b Small Finds 579 cp. -

A large, hooked cutting blade fragment measuring 56mm long and 32mm wide at the base of the blade. The socket and the point of the blade are missing. The sample was taken from the cutting edge.

Pit P545 Layer 3. Cunliffe, 1984a, Fig. 2.24.

D14a Small Finds 212 cp. 3

A tanged blade fragment measuring 72mm long, and 17mm wide at the base of the blade tapering to 4.5mm at the fracture. The point of the blade is missing. The sample was taken from the cutting edge near the tang.

Pit P107 Layer 2. Cunliffe, 1984a, Fig. 2.29.

D14b Small Finds 505 cp. 8

A small, hooked, cutting blade fragment measuring 57mm long, and 21mm wide at the base of the blade. A large portion of the point and socket of the blade are missing. The sample was taken from the cutting edge at the hook.

Layer 6. Cunliffe, 1984a, Fig. 2.9.

D15a Small Finds 559 cp. 7-8

A large, hooked, cutting blade fragment measuring 183mm long, and 45mm wide at the fracture. The point and the socket of the blade are missing. The sample was taken from the base of the blade.

Pit P511 Layer 3. Cunliffe, 1984a, Fig. 2.4.

D15b Small Finds 637 cp. 7

A tanged, bent, blade measuring 135mm long, and 36mm wide at the base of the blade tapering to a still sharp point. Two rivet holes pierce the tang. The sample was taken from the cutting edge.

Pit P595 Layer 2. Cunliffe, 1984a, Fig. 2.34.

D16a Small Finds 137 cp. 8

A small, socketed, tanged, hooked cutting blade fragment measuring 64mm long, and 19mm wide at the base of the blade. A portion of the point and the tang are missing. The sample was taken from the cutting edge at the hook.

Layer 6. Cunliffe, 1984a, Fig. 2.:6.

D16b Small Finds 960 cp. 6

A small, tanged, hooked, cutting blade measuring 0mm long, and 20mm wide at the base of the blade. Two riv ts pierce the tang. The sample was taken from the cuttin edge at the hook.

Pit P912 Layer 3. Cunliffe, 1984a, Fig. 2. .0.

D17 Small Finds 482 cp. 3

A tanged, hooked cutting blade measuring 140mm long, and 26mm wide at the base of the blade. The samp e was taken from the cutting edge at the hook.

Pit P428 Layer 3. Cunliffe, 1984a, Fig. 2.20.

D18 Small Finds 66 cp. 3

A tanged, palette shaped blade measuring 95mm l ng and 15mm wide at the base of the blade tapering to a rc nded point. The sample was taken from one cutting edge.

Trench 18 Layer 36. Cunliffe, 1984a, Fig. .33.

D19 Small Finds 362 cp. 7

A spiral finger ring measuring 23mm in diamete produced from a wire 7.5mm in diameter. The sample wa taken from one end of the wire.

Pit P329 Layer 1. Cunliffe, 1984a, Fig. 199.

D20 Small Finds 417 cp. 6

A strip fragment measuring 44.5mm long, 20mm w e, and 5mm thick. The sample was taken from one end.

Pit P321 Layer 5. Cunliffe, 1984a, Fig. 9 3.

D21a Small Finds 201 cp. 7

A roughly rectangular tang measuring 49mm long, 5mm wide, and 4mm thick. One end is fractured, and two r et holes pierce the tang. The base of a blade protrude from one side. The sample was taken from the side op site the protrusion.

Pit P79 Layer 4. Cunliffe, 1984a, Fig. 2 120.

D21b Small Finds 173 cp. 7

A rectangular strip measuring 98mm long, 18-20mm wide, and 2mm thick. A single perforation pierces the base of the strip. The sample was taken from one edge.

Pit P79 Layer 5. Cunliffe, 1984a, Fig. 2.113.

D22a Small Finds 96 cp. -

A folded strip measuring 81mm long, 14-20mm wide, and 2.5-4mm thick. The sample was taken from one edge.

Ph 126. Cunliffe, 1984a, Fig. 9:E10.

D22b Small Finds 205 cp. 7

A strip measuring 48mm long, 14mm wide, and 3mm thick. One end of the strip is bent and perforated. The sample was taken from one edge.

Pit P79 Layer 4. Cunliffe, 1984a, Fig. 2.114.

D23 Small Finds 177 cp. 7

A rod measuring 35mm long and 5mm in diameter. The sample was taken from one end.

Pit P79 Layer 5. Cunliffe, 1984a, Fig. 9:E12.

D24a Small Finds 774 cp. 6/8

A socketed object measuring 75mm long, 14-19mm wide, and 2.5mm thick. The artefact possesses a blade of undetermined function. The sample was taken from the blade just above the socket.

Pit P699 Layer 1. Cunliffe, 1984a, Fig. 9:F6.

D24b Small Finds 767 cp. 7

The artefact consists of two rectangular sheets measuring 43x16x1mm and 49x18x1mm in dimensions respectively, riveted together to form a length of 71mm. The sample was taken from the over-lap area near the rivet.

Pit P675 Layer 1. Cunliffe, 1984a, Fig. 2.110.

D25a Small Finds 426 cp. 7

A flat circular disc measuring 68.5mm in diameter with a central perforation 9mm in diameter. The sample was taken from the edge.

Pit P354 Layer 7. Cunliffe, 1984a, Fig. 2.174.

D25b Small Finds 837 cp. 3

A roughly rectangular strip fragment measuring 40mm long, 24mm wide, and 1.5mm thick. Two rivets are present towards the base of the piece. The sample was taken from the top edge.

Pit P816 Layer 1. Cunliffe, 1984a, Fig. 2.121.

D26a Small Finds 113 cp. -

A small wedge measuring 39mm long, and 8x5mm in dimensions below the deformed butt, tapering to a still sharp blade. The sample was taken from the blade.

Ph 476. Cunliffe, 1984a, Fig. 2.185.

D26b Small Finds 425 cp. 7

A bucket handle fragment measuring 381mm long, and 6mm in diameter. The looped attachments on each end of the handle are missing. The sample was taken from one end.

Pit P354 Layer 7. Cunliffe, 1984a, Fig. 2.171.

D27a Small Finds 171 cp. 7

A rectangular sheet measuring 64mm long, 21mm wide and 2mm thick. One edge is folded over, and a rivet hole pierces the centre of the artefact. The sample was taken from one end.

Pit P79 Layer 6. Cunliffe, 1984a, Fig. 2.124.

D27b Small Finds 1299 cp. 7

A clamp measuring 91mm long, 10.5mm wide, and 7-8mm thick. Both ends are bent in the same direction. The sample was taken from one end.

Layer 453. Cunliffe, 1984a, Fig. 2.155.

D28a Small Finds 295 cp. 7

A blade tang measuring 47mm long, 25mm wide and 2mm thick. Two rivet holes pierce the central axis. The sample was taken from one end.

Pit P26 Layer 3. Cunliffe, 1984a, Fig. 2.35.

D28b Small Finds 297 cp. 7

A broken rod measuring 70mm long, and 8x8mm in dimesnions. The sample was taken from the complete end.

Pit P26 Layer 6. Cunliffe, 1984a, Fig. 9:F1.

D29a Small Finds 119 cp. 6

A clamp produced by bending a 133mm long rod 4x8mm in dimensions, in two places. The sample was taken from one end.

Layer 8. Cunliffe, 1984a, Fig. 2.157.

D29b Small Finds 135 cp. 6

A small, curved rod measuring 131mm long, and 6mm in diameter. The artefact was probably a portion of a bucket handle. The sample was taken from one end.

Layer 8. Cunliffe, 1984a, Fig. 2.172.

D30a Small Finds 200 cp. 7

A roughly rectangular strip measuring 76.5mm long, 27-30mm wide, and 2mm thick. The sample was taken from one end.

Pit P79 Layer 4. Cunliffe, 1984a, Fig. 2.127.

D30b Small Finds 210 cp. 7

A large amorphous shaped ingot approximately 75mm long, 35mm wide, and 25mm thick. The sample was taken from one corner.

Pit P79 Layer 4. Cunliffe, 1984a, Fig. 2.205.

D31a Small Finds 742 cp. 7

A tanged object of undeterminable use measuring [illegible]0mm long, 6-9mm wide, and 5-7mm thick. The sample was taken from one edge above the tang.

Pit P652 Layer 2. Cunliffe, 1984a, Fig. 9 F6.

D31b Small Finds 1337 cp. 3?

A roughly triangular sheet fragment measuring approximately 40mm long on each side. A single perforation is present in one rounded corner, and portions of the other two corners are missing. The sample was taken from one edge

Ph3508 Layer 2. Cunliffe, 1984a, Fig. 2 125.

D32a Small Finds 139 cp. 3

A curved rod fragment measuring 74mm long, and 5mm in diameter at one end tapering to a sharp point at the other. The sample was taken from the pointed end.

Pit P63A Layer 6. Cunliffe, 1984a, Fig. 9:E11.

D32b Small Finds 824 cp. 1-3

A rectangular sheet measuring 40mm long, 20mm wide, and 3.5mm thick. The artefact has two large perforations along its central axis. The sample was taken from one end.

Layer 229. Cunliffe, 1984a, Fig. 2.141.

D33a Small Finds 134 cp. 5

A socket fragment measuring 48mm long, and 16-21mm wide. The sample was taken from one end.

Pit P75 Layer 1. Cunliffe, 1984a, Fig. 9:E11.

D33b Small Finds 1200 cp. 5

A rod fragment measuring 152mm long, and 4x4mm in dimensions at one end tapering to a sharp point at the other. The sample was taken from the pointed end.

Layer 374. Cunliffe, 1984a, Fig. 9:F10.

D101 Small Finds 14 cp. 7

A complete currency bar measuring 801mm long, 35mm wide, and 5mm thick. The sample was taken by C. Salter from the edge.

Trench 1 Layer 53. Cunliffe, 1984a, Fig. 2.77.

D102 Small Finds 15 cp. 7

A complete currency bar measuring 807mm long, 37mm wide, and 5mm thick. The sample was taken by C. Salter from the edge.

Trench 1 Layer 53. Cunliffe, 1984a, Fig. 2.78.

D103 Small Finds 16 cp. 7

A currency bar measuring 770mm long, 38mm wide, and 3mm thick. A portion of the socket and the shaft near the point are missing. The sample was taken by C. Salter from the edge.

Trench 1 Layer 53. Cunliffe, 1984a, Fig. 2.79.

D104 Small Finds 17 cp. 7

A currency bar measuring 810mm long, 36mm wide, and 4mm thick. A portion of the point and the end of the socket are missing. The sample was taken by C. Salter from the edge.

Trench 1 Layer 53. Cunliffe, 1984a, Fig. 2.80.

D105 Small Finds 18 cp. 7

A currency bar measuring 784mm long, 39mm wide, and 5mm thick. A small portion of the point and the end of the socket are missing. The sample was taken by C. Salter from the edge.

Trench 1 Layer 53. Cunliffe, 1984a, Fig. 2.81.

D106 Small Finds 19 cp. 7

A complete currency bar measuring 805mm long, 38mm wide, and 4mm thick. The sample was taken by C. Salter from the edge.

Trench 1 Layer 53. Cunliffe, 1984a, Fig. 2.82.

D107 Small Finds 20 cp. 7

A complete currency bar measuring 815mm long, 39mm wide, and 5mm thick. The sample was taken by C. Salter from the edge.

Trench 1 Layer 53. Cunliffe, 1984a, Fig. 2.83.

D108 Small Finds 21 cp. 7

A currency bar measuring 777mm long, 37mm wide, and 4mm thick. Portions of the socket are missing. The sample was taken by C. Salter from the edge.

Trench 1 Layer 53. Cunliffe, 1984a, Fig. 2.84.

D109 Small Finds 22 cp. 7

A currency bar measuring 799mm long, 35mm wide, and 3mm thick. Part of the point is missing. The sample was taken by C. Salter from the edge.

Trench 1 Layer 53. Cunliffe, 1984a, Fig. 2.85.

D110 Small Finds 23 cp. 7

A complete currency bar measuring 789mm long, 35mm wide, and 5mm thick. The sample was taken by C. Salter from the edge.

Trench 1 Layer 53. Cunliffe, 1984a, Fig. 2.86.

D111 Small Finds 24 cp. 7

A complete currency bar measuring 785mm long, 40mm wide, and 5mm thick. The sample was taken by C. Salter from the edge.

Trench 1 Layer 53. Cunliffe, 1984a, Fig. 2.87.

D112 Small Finds 25 cp. 7

A currency bar measuring 765mm long, 38mm wide, and 5mm thick. Part of the socket is missing. The sample was taken by C. Salter from the edge.

Trench 1 Layer 53. Cunliffe, 1984a, Fig. 2.88.

D120 Small Finds 207 cp. 7

A socketed tool fragment measuring 67mm long, 16mm wide at the top of the socket, and 6-7.5mm thick. The base of the socket, and a flattened tool of undetermined use are missing. The sample was taken by C. Salter from the edge of the top of the socket.

Pit P79 Layer 4. Cunliffe, 1984a, Fig. 9:E13.

D121 Small Finds 2114 cp. -

A blade point measuring 32mm long, and 13mm wide at the fracture tapering to a sharp point. The sample was taken by C. Salter from the cutting edge near the point.

Layer 393. Cunliffe, 1984a, Fig. 2.38.

D122 Small Finds 1203 cp. 7

A currency bar point measuring 104mm long, 28mm wide, and 4mm thick. The sample was taken by C. Salter from the edge.

Layer 393. Cunliffe, 1984a, Fig. 2.99.

D123 Small Finds 209 cp. 7

A complete, tanged object, of undetermined use, measuring 96mm long, 28mm wide at the top of the tang, and 20mm thick. The sample was taken by C. Salter from the edge at the top of the tang.

Pit P79 Layer 4. Cunliffe, 1984a, Fig. 9:E14.

D124 Small Finds 702 cp. 4

A clamp measuring 125mm long, 4.5mm wide, and 5mm thick. The artefact is bent in three places to form a rectangular outline. The sample was taken by C. Salter from the centre section of the clamp.

Pit P617 Layer 2. Cunliffe, 1984a, Fig. 9:F5.

D125 Small Finds 1205 cp. -

A rod fragment measuring 49mm long, and 5x5mm in dimensions at one end tapering to 4x4mm at the other. The sample was taken by C. Salter from one edge.

Layer 393. Cunliffe, 1984a, Fig. 9:F10.

D126 Small Finds 1211 cp. -

A nail measuring 17mm long, with a forged flattened head. The sample was taken by C. Salter from the head.

Layer 393. Cunliffe, 1984a, Fig. 2.183.

D127 Small Finds 175 cp. 7

A bar fragment measuring 51mm long, and 15x7mm in dimensions. The sample was taken by C. Salter from one edge near one end.

Pit P79 Layer 5. Cunliffe, 1984a, Fig. 9:E12.

D128 Small Finds 739 cp. 7

A complete ring staple measuring 45mm long. The artefact was produced by forging a rod measuring 5-6mm in diameter into a ring at one end. The sample was taken by C. Salter from the back of the ring.

Pit P672 Layer 2. Cunliffe, 1984a, Fig. 2.188.

D129 Small Finds 208 cp. 7

A socketed ferrule measuring 57.5mm long, and 20mm wide at the top of the socket tapering to a point. The sample was taken by C. Salter from the socket.

Pit P79 Layer 4. Cunliffe, 1984a, Fig. 2.52.

D130 Small Finds 211 cp. 7

A small, hooked cutting blade measuring 81mm long, and 22mm wide at the base of the blade. A single rivet pierces the base of the socket between the two flanges. The sample was taken by C. Salter from the cutting edge at the hook.

Pit P79 Layer 4. Cunliffe, 1984a, Fig. 2.5.

D131 Small Finds 1238 cp. 7

A complete tool of undetermined use measuring 122mm long, 9mm wide and 6.5mm thick. The bar has been bent to a right angle near the point, and flattened at the tip. The sample was taken by C. Salter from one side.

Pit P978 Layer 1. Cunliffe, 1984a, Fig. 2.68.

D132 Small Finds 750 cp. 7

A rod fragment measuring 44mm long and 7mm in diameter. The sample was taken by C. Salter from one end.

Pit P672 Layer 2. Cunliffe, 1984a, Fig. 9:F6.

D133 Small Finds 1227 cp. 7

A tanged file measuring 68mm long, 8mm wide at the base of the blade tapering to 3mm wide at the fracture, and 2mm thick. The bottom of the tang and the tip of the file are missing. The sample was taken by C. Salter from one side of the artefact near the fracture.

Layer 393. Cunliffe, 1984a, Fig. 2.55.

D134 Small Finds 1227 cp. 7

A second sample was taken from the artefact described in D133. The sample was taken by C. Salter from the edge of the artefact near the tip.

Layer 393. Cunliffe, 1984a, Fig. 2.55.

D135 Small Finds 13 cp. 7

A heavy blade fragment measuring 66mm long, and 16-29mm wide. The sample was taken by C. Salter from the cutting edge.

Trench 9 Layer 8. Cunliffe, 1984a, Fig. 2.41.

D136 Small Finds 336 cp. 1-3

A currency bar measuring 115mm long, 36mm wide at the top of the socket, and 4mm thick. Most of the blade is missing. The sample was taken by C. Salter from the edge.

Pit P149 Layer 2. Cunliffe, 1984a, Fig. 2.97.

D137 Small Finds 388 cp. 7

A socketed ferrule measuring 107mm long, and 22mm wide at the base of the socket tapering to 7mm at the tip. A substantial portion of the socket is missing. The sample was taken by C. Salter from one edge near the tip.

Pit P286 Layer 10. Cunliffe, 1984a, Fig. 2.53.

D138 Small Finds 541 cp. 6

A second sample was taken from the artefact described in D12. The sample was taken by C. Salter from the loop.

Layer 4. Cunliffe, 1984a, Fig. 2.165.

D139 Small Finds 50 cp. 5

A socketed chisel measuring 176mm long, and 23mm in diameter at the socket tapering to 13mm at the head. Portions of the socket and the blade are missing. The sample was taken by C. Salter from the blade.

Trench 11 Layer 68. Cunliffe, 1984a, Fig. 2.46.

D140 Small Finds 49 cp. -

A lobate horse shoe fragment measuring 94mm long. Two nails and three punctures are present. The sample was taken by C. Salter from the edge near the intact end.

Trench 12 Layer 14. Cunliffe, 1984a, Fig. 2.73.

D141 Small Finds 227 cp. 7

A tang fragment measuring 40mm long, and 6x8mm in dimensions. An oval washer is attached at one end, and measures 34mm long, 27mm wide, and 2mm thick. The sample was taken by C. Salter from the tang at the end not possessing the washer.

Pit P110 Layer 1. Cunliffe, 1984a, Fig. 2.175.

D142 Small Finds 968 cp. 7

A blade measuring 73mm long, and 17mm wide at the base of the blade tapering to 11.5mm at the fracture. Two rivets pierce the base, and the point is missing. The sample was taken by C. Salter from the cutting edge near the rivets.

Pit P878 Layer 7. Cunliffe, 1984a, Fig. 2.36.

D143 Small Finds 95 cp. 5

A ring measuring 17.5mm in diameter, and produced from a rod measuring 20mm wide and 3.5mm thick. A small tang of dimensions 6x8mm, protrudes from one edge. The sample was taken by C. Salter from one edge of the ring.

Pit P32B Layer 2. Cunliffe, 1984a, Fig. 2.142.

D144 Small Finds 203 cp. 7

A bar fragment measuring 44mm long, and 12.5x12.5mm in dimensions at one end tapering to 10x7.5mm at the other. The sample was taken by C. Salter from one edge near the thick end.

Pit P79 Layer 4. Cunliffe, 1984a, Fig. 9:E13.

D145 Small Finds 539 cp. 7

A bar fragment measuring 34mm long, 27-33mm wide, and 7.5-8.5mm thick. The sample was taken by C. Salter from one end.

P433 Layer 5. Cunliffe, 1984a, Fig. 9:F4.

D146 Small Finds 935 cp. -

A tanged blade measuring 124mm long, 12mm wide at the base of the blade tapering to a sharp point, and 3mm thick. The sample was taken by C. Salter from the cutting edge near the tang.

Unstratified. Cunliffe, 1984a, Fig. 2.28.

D147 Small Finds 932 cp. 7

A tanged blade fragment measuring 82mm long, and 26mm wide at the base of the blade. The bottom of the tang and the point of the blade are missing. The sample was taken by C. Salter from the tang.

Pit P896 Layer 3. Cunliffe, 1984a, Fig. 2.27.

D148 Small Finds 304a cp. -

A nail measuring 50mm long, with a forged flattened head. The sample was taken by C. Salter from the head.

Unstratified. Cunliffe, 1984a, Fig. 9:F2.

D149 Small Finds 304b cp. -

A nail measuring 50mm long with a forged flattened head. The sample was taken by C. Salter from the head.

Unstratified. Cunliffe, 1984a, Fig. 9:F2.

D150 Small Finds 304c cp. -

A nail fragment measuring 42.5mm long with a forged flattened head. The sample was taken by C. Salter from the head.

Unstratified. Cunliffe, 1984a, Fig. 9:F2.

D151 Small Finds 676 cp. 7

A round-headed wedge measuring 62mm long, and 10x12mm in dimensions below the head tapering to 7x7.5mm at the fracture. The base of the wedge is missing. The sample was taken by C. Salter near the fracture.

Pit P556 Layer 4. Cunliffe, 1984a, Fig. 2.186.

D152 Small Finds 174 cp. 7

A z-shaped object produced from a bar measuring 70.5mm long, 4-9mm wide, and 4-5mm thick. The sample was taken by C. Salter from the middle section.

Pit P79 Layer 5. Cunliffe, 1984a, Fig. 2.164.

D153 Small Finds 818 cp. 7

A tanged object measuring 85mm long and 45x5x4mm at the tang. The broken blade measures 13x4mm in dimensions. The sample was taken by C. Salter from the fractured end.

Layer 214. Cunliffe, 1984a, Fig. 9:F7.

D154 Small Finds 1302 cp. 3

A tanged, socketed, hooked cutting blade measuring 118mm long, and 34mm wide at the base of the blade. The point of the blade and the base of the tang are missing. A rivet hole pierces the base of the socket between the two flanges. The sample was taken by C. Salter from the cutting edge near the socket.

Pit P1034 Layer 1. Cunliffe, 1984a, Fig. 2.3.

D155 Small Finds 1326 cp. 4-5

A roughly rectangular bar fragment measuring 48mm long, 37mm wide, and 7mm thick. The sample was taken by C. Salter from one edge.

Layer 425. Cunliffe, 1984a, Fig. 9:F14.

D156 Small Finds 1326 cp. 4-5

A second sample was taken from the artefact described in D155. The sample was taken by C. Salter from the opposite end.

Layer 425. Cunliffe, 1984a, Fig. 9:F14.

D157 Small Finds 1346 cp. 7

A chisel fragment measuring 63mm long, and 6x6mm in dimensions at the tang tapering to 12x2.5mm at the fracture. Portions of the blade and the tang are missing. The sample was taken by C. Salter from the fracture.

Layer 522. Cunliffe, 1984a, Fig. 2.48.

D158 Small Finds 1346 cp. 7

A second sample was taken from the artefact described in D157. The sample was taken by C. Salter near the fracture.

Layer 522. Cunliffe, 1984a, Fig. 2.48.

D159 Small Finds 1372 cp. 5

A roughly rectangular bar fragment measuring 32mm long, 28mm wide, and 5-5.5mm thick. The sample was taken by C. Salter from one end.

Pit P1054 Layer 1. Cunliffe, 1984a, Fig. 9:F14.

D160 Small Finds 974 cp. 7

A rectangular strip measuring 75mm long, 16mm wide, and 1.5mm thick. The sample was taken by C. Salter from one end.

Pit P911 Layer 6. Cunliffe, 1984a, Fig. 2.128.

D161 Small Finds 44 cp. -

A blade point measuring 55mm long and 28mm wide at the fracture tapering to a rounded point. The sample was taken by C. Salter from the cutting edge near the fracture.

Trench 14 Layer 20. Cunliffe, 1984a, Fig. 2.40.

D162 Small Finds 44 cp. -

A second sample was taken from the artefact described in D161. The sample was taken by C. Salter from the back edge of the blade near the point.

Trench 14 Layer 20. Cunliffe, 1984a, Fig. 2.40.

D163 Small Finds 441 cp. 7

A socketed ploughshare measuring 165mm long, and 42x15.5mm in dimensions at the top of the socket tapering to 30x8mm at the rounded point. The sample was taken by C. Salter from the point.

Pit P354 Layer 7. Cunliffe, 1984a, Fig. 2.70.

D164 Small Finds 638 cp. 7

An adze head measuring 133mm long, and 54x4mm in dimensions at the cutting edge tapering to 20x21mm at the socket. The sample was taken by C. Salter from the cutting edge.

Pit P589 Layer 10. Cunliffe, 1984a, Fig. 2.49.

D165 Small Finds 838 cp.7

A tanged, socketed, hooked cutting blade measuring 94mm long, and 29mm wide at the base of the blade. The tip of the blade and part of the tang are missing. The sample was taken by C. Salter from the cutting edge at the hook.

Layer 230. Cunliffe, 1984a, Fig. 2.14.

D166 Small Finds 953 cp. 7

A hammer, adze or pick socket and butt fragment measuring 51mm long and 20mm thick. The sample was taken by C. Salter from the butt.

Pit P911 Layer 1. Cunliffe, 1984a, Fig. 2.51.

D167 Small Finds 953 cp. 7

A second sample was taken from the artefact described in D166. The sample was taken by C. Salter from the fractured socket.

Pit P911 Layer 1. Cunliffe, 1984a, Fig. 2.51.

D168 Small Finds 833 cp. 6

A large, socketed pick measuring 210mm long, and 54.5x50mm in dimensions at the top of the socket tapering to 10.5x25mm at the point. The sample was taken by C. Salter from one side near the point.

Pit P810 Layer 1. Cunliffe, 1984a, Fig. 2.50.

D169 Small Finds 833 cp. 6

A second sample was taken from the artefact described in D168. The sample was taken by C. Salter from one side near the socket.

Pit P810 Layer 1. Cunliffe, 1984a, Fig. 2.50.

D170 Small Finds 833 cp. 6

A third sample was taken from the artefact described in D168. The sample was taken by C. Salter from the base of the socket.

Pit P810 Layer 1. Cunliffe, 1984a, Fig. 2.50.

D171 Small Finds 833 cp. 6

A fourth sample was taken from the artefact described in D168. The sample was taken by C. Salter from the base of the socket.

Pit P810 Layer 1. Cunliffe, 1984a, Fig. 2.50.

D172 Small Finds 146 cp. 7

A tanged file measuring 129mm long, 10-11mm wide, and 2mm thick. The teeth are set 4mm apart. The sample was taken by C. Salter from one edge of the blade.

Pit P27 Layer 4. Cunliffe, 1984a, Fig. 2.54.

D173 Small Finds 1396 cp. 8

A tanged saw blade measuring 353mm long, 40mm wide at the top of the tang tapering to 18mm wide at the tip, and 3-4mm thick. The teeth are partially set to cut on the pull stroke. The saw was broken into two pieces and later repaired in the Iron Age by lapwelding the sections together. Two rivets pierce the tang. One of the teeth was removed by C. Salter for examination.

Pit P1078 Layer 10. Cunliffe, 1984a, Fig. 2.42.

D174 Small Finds 1396 cp. 8

A second sample was taken from the artefact described in D173. The sample was taken by C. Salter from the back of the blade.

Pit P1078 Layer 10. Cunliffe, 1984a, Fig. 2.42.

D175 Small Finds 949 cp. 7

A rod measuring 47mm long, and 4.6x5mm in dimensions at one end tapering to 2x2mm at the other. The sample was taken by C. Salter from one edge near the wider end.

Pit P878 Layer 4. Cunliffe, 1984a, Fig. 9:F9.

D176 Small Finds 952 cp. 7

A tanged chisel measuring 136mm long, and 3mm wide at the blade. The sample was taken by C. Salter from the side of the artefact.

Pit P878 Layer 5. Cunliffe, 1984a, Fig. 2.60.

D177 Small Finds 1235 cp. 3

A bar fragment measuring 37mm long, 36mm wide, and 6.5mm thick. The sample was taken by C. Salter from one end.

Pit P982 Layer 2. Cunliffe, 1984a, Fig. 9:F11.

D178 Small Finds 1520 cp. -

An object consisting of a 230mm long z-shaped shaft with a ring at one end. The artefact was produced from a rod measuring 8mm in diameter. The sample was taken by C. Salter from the shaft.

F68 Layer 7. Unpublished.

D179 Small Finds 1517 cp. -

A rectangular plate fragment measuring 66mm long, 14mm wide, and 6-7mm thick. The sample was taken by C. Salter from one end.

F68. Unpublished

D180 Small Finds 1548 cp. -

A rod measuring 84mm long, and 3mm in diameter. The sample was taken by C. Salter from one side near an end.

Pit P1385 Layer 1. Unpublished.

D181 Small Finds 1525 cp. -

A rod fragment measuring 81mm long, and 10x10mm in dimensions at one end tapering to a flattened, split 13x3mm opposing end. The sample was taken by C. Salter from the complete end.

Pit P1349 Layer 1. Unpublished.

D182 Small Finds 1535 cp. -

A blade fragment measuring 67mm long and 19-22mm wide. The sample was taken by C. Salter from one side.

Pit P1345 Layer 3. Unpublished.

D183 Small Finds 1519 cp. -

A rectangular sheet measuring 130mm long, 33mm wide at one end tapering to 13mm at the other, and 5-6mm thick. The sample was taken by C. Salter from one edge.

F68 Layer 6. Unpublished.

D184 . Small Finds 1585 cp. -

A clamp fragment measuring 42mm long, 6mm wide, and 2-5mm thick. The artefact is bent in two places to form three side of a rectangle. The sample was taken by C. Salter from the centre section.

Layer 613. Unpublished.

D185 Small Finds 1522 cp. -

A wedge measuring 130mm long with a forged flattened head. The sample was taken by C. Salter from one side towards the middle of the piece.

F68 Layer 9. Unpublished.

D186 Small Finds 1522 cp. -

A second sample was taken from the artefact described in D185. The sample was taken by C. Salter from the blade.

F68 Layer 9. Unpublished.

D187 Small Finds 1515 cp. -

A cauldron hook measuring 282mm long. The sample was taken by C. Salter from the shaft near the hook.

F68 Layer 2. Unpublished.

D188 Small Finds 1515 cp. -

A second sample was taken from the artefact described in D187. The sample was taken by C. Salter from the hook.

F68 Layer 2. Unpublished.

D189 Small Finds 1521 cp. -

A curved rod measuring 239mm long, and 7mm in diameter. The sample was taken by C. Salter from the complete end.

F68 Layer 8. Unpublished.

FBD1a cp. 3/4

A small, socketed, tanged, hooked cutting blade measuring 61mm long, and 19mm wide at the base of the blade. The sample was taken from the cutting edge between the point and the hook.

Pit 27. Clay, 1924, Fig. 11-4.

FBD1b cp. 3/4

A large, tanged blade measuring 218mm long, and 34mm wide at the base of the blade tapering to a slightly hooked, rounded point. The sample was taken from the cutting edge.

Pit 18. Clay, 1924, Fig. 11-1.

HNY1a Museum Acc. D.163/1957-8 cp. -

A socketed spearhead measuring 102mm long, and 20mm wide just above the socket. Most of the blade is missing. The sample was taken from the cutting edge near the fracture.

HNY1b Museum Acc. D.355/1956-7 cp. -

A socketed spearhead measuring 122mm long, and 25mm wide just above the socket. The point of the blade is missing. The sample was taken from one edge near the fracture.

HNY2a Museum Acc. D.352/1956-7 cp. -

A socketed spearhead measuring 87mm long, and 16mm wide just above the socket. The point and the edges of the blade are missing. A small midrib is present. The sample was taken from the edge of the blade just above the socket.

HNY2b Museum Acc. D.293/1956-7 cp. -

A tanged dagger measuring 192mm long, and 25mm wide just above the tang tapering to 13mm wide at the rounded point. The sample was taken from one edge of the blade near the tang.

HNY3a Museum Acc. D.346/1956-7 cp. -

A socketed spearhead measuring 124mm long, and 15mm wide just above the socket. Most of the point and the edges of the blade are missing. The sample was taken from the edge of the blade just below the tip.

HNY3b Museum Acc. D.347/1956-7 cp. -

A socketed spearhead measuring 122mm long, and 26mm wide. The point of the blade is missing. The sample was taken from one edge of the blade.

HNY4a Museum Acc. D.344/1956-7 cp. -

A socketed spearhead measuring 169mm long, and 14mm wide from the midrib to the remaining corroded edge. One side of the blade and half of the other side are missing. The sample was taken from one of the corroded edges.

HNY4b Museum Acc. D.320/1956-7 cp. -

An adze measuring 174mm long, 48mm wide at the edge of the blade tapering to 21mm wide just above the socket, and 28mm thick just above the socket tapering to a still sharp edge. The sample was taken from the edge.

HNY5a Museum Acc. D.356/1956-7 cp. -

A midribbed blade, spearhead, or dagger fragment measuring 103mm long and 33mm wide at the midsection tapering to 20mm at the fracture The point and the tang or socket of the object are missing. The sample was taken from the edge of the blade.

HNY5b Museum Acc. D.130/1957-8 cp. -

A large, curved strip fragment measuring 166mm long, 32mm wide and 10mm thick. The artefact was produced from two 5mm. thick strips. The sample was taken from the edge at the maximum width.

HNY6a Museum Acc. D.121/1957-8 cp. -

A currency bar handle measuring 120mm long, 33mm wide at the top of the socket, and 5mm thick. The sample was taken from one edge.

HNY6b Museum Acc. D.116/1957-8 cp. -

A curved strip measuring 194mm long, 33mm wide, and 4mm thick. A mass of iron measuring 19mm thick and approximately 27mm in diameter is present at one end. The sample was taken from the flat end.

HNY7a Museum Acc. D.132/1957-8 cp. -

A currency bar blade fragment measuring 179mm long, 37mm wide, and 6mm thick. The artefact was produced by incompletely welding three plates laterally together. The sample was taken from one edge.

HNY7b Museum Acc. D.366D/1956-7 cp. -

A chariot tyre fragment measuring 192mm long, 29mm wide, and 4mm thick. The sample was taken from one edge.

HNY8a Museum Acc. D.134/1957-8 cp. -

A currency bar handle measuring 136mm long, 44mm wide, and 5-9mm thick. The plates forming the object are incompletely welded together. The sample was taken from the flat end.

HNY8b Museum Acc. D.120/1957-8 cp. -

A currency bar blade measuring 323mm long, 35mm wide, and 4mm thick. The artefact is fractured at both ends. The sample was taken from one edge.

HNY9a Museum Acc. D.119/1957-8 cp. -

A currency bar handle fragment measuring 323mm long, 29mm wide, and 4mm thick. The point of the artefact is missing. The sample was taken from one edge near the incomplete socket.

HNY9b Museum Acc. D.379/1956-7 cp. -

A currency bar blade fragment measuring 176mm long, 28-32mm wide, and 3-4mm thick. The sample was taken from one edge.

HNY10a Museum Acc. D.139/1957-8 cp. -

A ploughshare point measuring 180mm long, 25mm wide at the fracture tapering slightly to a blunt point, and 9mm thick at the fracture tapering to 2mm at the point. The sample was taken from the point.

HNY10b Museum Acc. D.333/1956-7 cp. -

A chisel measuring 179mm long, 14mm wide, and 14mm thick at the butt tapering to a still sharp blade. The butt of the chisel possesses a burr from being struck with a hammer. The sample was taken from the blade.

HNY11a Museum Acc. D.342/1956-7 cp. -

A socketed spearhead measuring 225mm long, and 20mm wide at the top of the socket tapering to a point. The midrib measures 7mm thick at the socket and tapers to the point. The sample was taken from one edge.

HNY11b Museum Acc. D.340/1956-7 cp. -

A spearhead point fragment measuring 149mm long, and 20mm wide one end tapering to a blunt point. The midrib measures 8mm thick at the fracture. The sample was taken from the edge near the fracture.

HNY12a Museum Acc. D.351.1956-7 cp. -

A socketed spearhead measuring 168mm long, and 25mm wide at the widest portion of the blade. One side of the blade and the point are missing. The midrib is 9mm thick at the top of the socket and tapers to the fracture. The sample was taken from one edge of the blade.

HNY12b Museum Acc. D.361/1956-7 cp. -

A socketed ploughshare measuring 106mm long, 40mm wide at the base of the socket tapering to a blunt point, and 2mm thick at the point. A large portion of the socket is missing. The sample was taken from the point.

HNY13a Museum Acc. D.130/1957-8 cp. -

A currency bar blade fragment measuring 136mm long, 35mm wide, and 4-6mm thick. The sample was taken from one edge.

HNY13b Museum Acc. D.354/1956-7 cp. -

A socketed spearhead measuring 187mm long, and 19mm wide at the base of the socket. The blades on both sides of the midrib are missing. The midrib is 8mm thick at the socket, tapering to the point. A single rivet hole pierces the socket. The sample was taken from the corroded edge of the blade.

HNY14a Museum Acc. D.341/1956-7 cp. -

A socketed spearhead measuring 154mm long, and 18mm wide at the base of the socket. The edges of the blades are missing. The sample was taken from the corroded edge of the blade.

HNY14b Museum Acc. D.359/1956-7 cp. -

A ploughshare measuring 134mm long, and 35x5mm in dimensions at the middle of the socket tapering to 14x5mm at the rounded point. One side of the socket is missing. The sample was taken from the edge just behind the point.

HNY15a Museum Acc. D.123/1957-8 cp. -

A socketed ploughshare measuring 443mm long, 28mm wide just above the socket tapering to a blunt point, and 8mm thick. The blade of the ploughshare was produced by incompletely welding three iron sheets laterally together. The sample was taken from the point.

HNY15b Museum Acc. D.292/1956-7 cp. -

A tanged dagger fragment measuring 333mm long, and 15mm wide at the fracture just below the point. The point and the edges of the blade are missing. The sample was taken from one edge of the blade.

HNY16a Museum Acc. D.122/1957-8 cp. -

A currency bar blade fragment measuring 418mm long, 32mm wide at one end tapering to 19mm at the other, and 4mm thick. The sample was taken from one edge.

HNY16b Museum Acc. D.117/1957-8 cp. -

A currency bar point fragment measuring 215mm long, and 36mm wide at the fracture tapering to 20mm just above the corroded point. The sample was taken from the edge.

HNY17a Museum Acc. D.294/1956-7 cp. -

A tanged dagger measuring 349mm long, and 40mm wide at the base of the blade tapering to a blunt point. The sample was taken from the blade.

HNY17b Museum Acc. D.343/1956-7 cp. -

A large, socketed spearhead measuring 353mm long, and 42mm wide at the base of the blade tapering to a blunt point. The midrib is 9mm thick at the socket and also tapers to the point. The sample was taken from one edge of the blade.

HNY18a Museum Acc. D.332/1956-7 cp. -

A chisel measuring 226mm long, 12mm wide, and 12mm thick at the butt end tapering to a still sharp blade 12mm wide. There is no evidence of a burr on the butt end. The sample was taken from the blade.

HNY18b Museum Acc. D.390/1956-7 cp. -

A blacksmith's poker consisting of a 293mm long, 11mm diameter shaft attached to a flat, ovally shaped head measuring 120mm long, 36mm wide, and 6-8mm thick. The sample was taken from the tip of the disc.

HNY19a Museum Acc. D.392/1956-7 cp. -

A blacksmith's poker consisting of a 322mm long, 11mm diameter shaft attached to a flat, ovally shaped head measuring 90mm long, 52mm wide, and 3-5mm thick. The sample was taken from the tip of the disc.

HNY19b Museum Acc. D.339/1956-7 cp. -

A socketed spearhead measuring 214mm long, and 28mm wide at the base of the blade tapering to 10mm at the rounded point. The midrib also tapers from 8mm thick just above the socket to the point. A single rivet pierces the socket. The sample was taken from the blade.

HNY20a Museum Acc. D.336/1956-7 cp. -

A small pick measuring 132mm long, 10-11mm wide, and 8mm thick at the socket tapering to a still sharp edge. The butt of the artefact possesses a burr from being struck with a hammer. The sample was taken from the blade.

HNY20b Museum Acc. D.324/1956-7 cp. -

A small, hooked cutting blade measuring 140mm long, and 15mm wide at the fracture. The tang or socket of the blade is missing. The sample was taken from the cutting edge at the hook.

HNY21a Museum Acc. D.334/1956-7 cp. -

A chisel measuring 202mm long, 12mm wide, and 5mm thick at the butt tapering to a sharp blade. The butt of the chisel possesses a burr from being struck with a hammer. The sample was taken from the blade.

HNY21b Museum Acc. D.296/1956-7 cp. -

A chisel measuring 138mm long, 8-9mm wide at the butt flaring to 21mm at the blade, and 4mm thick. A simple iron hilt is present 53mm down the shaft from the base. The sample was taken from the blade.

HNY22a Museum Acc. D.349/1956-7 cp. -

A large, leaf-shaped, socketed spearhead measuring 169mm long, and 28mm wide from the midrib to the complete side of the blade, tapering to 4mm wide at the blunt point. One side of the blade is missing. The sample was taken from the complete side of the blade near the point.

HNY22b Museum Acc. D.310/1956-7 cp. -

A tanged blade measuring 156mm long, and 35mm wide at the base of the blade tapering to 13mm wide at the rounded point. The sample was taken from the cutting edge.

HNY23a Museum Acc. D.118/1957-8 cp. -

A currency bar point measuring 23mm long, 34mm wide, and 4mm thick. The artefact is bent at the fracture. The sample was taken from one edge near the point.

HNY23b Museum Acc. D.331/1956-7 cp. -

A chisel measuring 279mm long, 9-10mm wide, and 9-10mm thick tapering to a sharp blade. The sample was taken from the blade.

HNY24a Museum Acc. D.387/1956-7 cp. -

An awl shaped rod measuring 108mm long, and 10x10mm in dimensions at one end tapering to 8x10mm just before the 4mm wide blade. The sample was taken from the blade.

HNY24b Museum Acc. D.388/1956-7 cp. -

An awl shaped rod measuring 86mm long, and 5x5mm in dimensions at the blunt end flaring to 8x8mm in the middle, and then tapering to a 5mm wide sharp edge. The sample was taken from the sharp edge.

HNY25a Museum Acc. D.305/1956-7 cp. -

A tanged blade measuring 100mm long, and 24mm wide at the base of the blade tapering to 7mm wide at the fracture. The point of the blade is missing. The sample was taken from the cutting edge.

HNY25b Museum Acc. D.366/1956-7 cp. -

A chariot tyre fragment measuring 324mm long, 36mm wide, and 4-5mm thick. The sample was taken from one edge.

HNY26a Museum Acc. D.353/1956-7 cp. -

A spearhead fragment measuring 214mm long, and 17mm wide. The 8mm thick midrib tapers to the point. Neither side of the blade is present. The sample was taken from the midrib.

HNY26b Museum Acc. D.348/1956-7 cp. -

A socketed, leaf-shaped spearhead measuring 132mm long, and 31mm wide at the base of the blade tapering to 9mm wide at the rounded point. The sample was taken from the blade near the top of the socket.

HNY27a Museum Acc. D.380/1956-7 cp. -

A pick measuring 128mm long, 9mm wide, and 12mm thick at the socket tapering to a still sharp edge. The blade was produced by incompletely welding two iron sheets laterally together. The sample was taken from both sheets of the blade.

HNY27b Museum Acc. D.164/1957-8 cp. -

A socketed, bent spearhead measuring 217mm long, and 34mm wide. The edges of the blade and the point are missing. The sample was taken from one edge of the blade.

HNY28a Museum Acc. D.295/1956-7 cp. -

A tanged dagger measuring 298mm long, and 38mm wide at the base of the blade tapering to a blunt point. There is no midrib. The sample was taken from a cutting edge near the point.

HNY28b Museum Acc. D.295/1956-7 cp. -

A saw fragment measuring 118mm long, 30mm wide and 3mm thick. The teeth are not aligned. This artefact is part of a complete saw with tang and bone handle. One of the teeth was removed for examination.

HNY29a Museum Acc. D.357/1956-7 cp. -

A socketed, hooked cutting blade measuring 121mm long, and 28mm wide at the base of the blade. A section of the blade is missing above the hook. The sample was taken from the cutting edge below the fracture.

HNY29b Museum Acc. D.327/1956-7 cp. -

A small, hooked, cutting blade measuring 91mm long, and 21mm wide at the base of the blade. One rivet hole pierces the base. The base curves slightly, which may indicate a socket once existed. The sample was taken from the cutting edge at the hook.

HNY30a Museum Acc. D.291/1956-7 cp. -

A tanged dagger measuring 185mm long, 33mm wide at the base of the blade, and 10mm thick along the central axis. The artefact is fractured 80mm up the blade. The sample was taken from the cutting edge near the fracture.

HNY30b Museum Acc. D.288/1956-7 cp. -

A sword scabbard fragment measuring 63-68mm long, 35-37mm wide, and 7mm thick at the edge where it would have been connected to the top half of the scabbard, but 2mm thick predominantly. The sample was taken from the thinner edge.

HNY31a Museum Acc. D.301/1956-7 cp. -

A tanged, bent blade measuring 145mm long, and 25mm wide. The sample was taken from the cutting edge.

HNY31b Museum Acc. D.362/1956-7 cp. -

A linch-pin measuring 144mm long. The sample was taken from the top terminal.

HNY32a Museum Acc. D.315/1956-7 cp. -

A tanged blade measuring 129mm long, and 29-31mm wide at the base of the blade. The artefact is bent at a right angle just above the tang. The point of the blade is missing. The sample was taken from the cutting edge near the fracture.

HNY32b Museum Acc. D168/1957-8 cp. -

A hooked, cutting blade fragment measuring 102mm long, and 38mm wide at the fracture tapering to a blunt point. The socket of the blade is missing. The sample was taken from the cutting edge.

HNY33a Museum Acc. D.318/1956-7 cp. -

A large hammer measuring 186mm long, 40mm wide just in front of the socket tapering to 8mm at the edge, and 42mm thick at the socket flaring to 61mm at the edge. The sample was taken from the edge.

HNY33b Museum Acc. D.300/1956-7 cp. -

A tanged blade measuring 120mm long, and 32mm wide at the base of the blade tapering to 22mm wide at the fracture. The point of the blade is missing. One rivet pierces the tang. The sample was taken from the cutting edge.

HNY34a Museum Acc. D.308/1956-7 cp. -

A small, tanged blade measuring 109mm long, and 19mm wide at the base of the blade tapering to 9mm at the tip. The sample was taken from the cutting edge.

HNY34b Museum Acc. D.298/1956-7 cp. -

A tanged, bent blade measuring 195mm long, and 29mm wide at the base of the blade tapering to 19mm just before the point. The sample was taken from the cutting edge near the base of the blade.

HNY35a Museum Acc. D.356/1956-7 cp. -

A hooked strip measuring 82mm long, 21-23mm wide, and 4mm thick. The sample was taken from the exterior edge.

HNY35b Museum Acc. D.389/1956-7 cp. -

A sword measuring 782mm long, and 47mm wide at the base of the blade tapering to 26mm wide at the point. The bronze chape is still intact. The sample was taken from one edge of the sword.

HNY36a Museum Acc. D.165/1957-8 (1) cp. -

A small, tanged, slightly hooked blade measuring 95mm long, and 19mm wide at the base of the blade. The sample was taken from the cutting edge near the point.

HNY36b Museum Acc. D.297/1956-7 cp. -

A tanged blade measuring 130mm long, and 31mm wide at the base of the blade tapering to 13mm wide at the rounded point. The tang is only 21mm long and terminates in a flat stump 10mm in diameter. The sample was taken from the cutting edge near the base of the blade.

HNY37a Museum Acc. D.153/1957-8 (2) cp. -

A blade fragment measuring 103mm long, and 25mm wide at the base of the blade tapering to 13mm at the fracture. The tang and the point are missing. The sample was taken from the cutting edge.

HNY37b Museum Acc. D.312/1956-7 cp. -

A tanged blade measuring 132mm long and 20mm wide at the base of the blade tapering to 11mm wide at the rounded point. The tang is only 4mm long and terminates in a flat stump 10mm in diameter. The sample was taken from the cutting edge.

HNY38a Museum Acc. D.313/1956-7 cp. -

A short, tanged blade measuring 100mm long, and 23mm wide at the base of the blade tapering to 8mm at the rounded point. A rivet pierces the tang. The sample was taken from the cutting edge near the base.

HNY38b Museum Acc. D.325/1956-7 cp. -

A hooked blade fragment measuring 106mm long, and 28mm wide at the fracture at the base. The socket and the point of the blade are missing. The sample was taken from the cutting edge.

HNY39a Museum Acc. D.326/1956-7 cp. -

A hooked blade measuring 118mm long and 29mm wide at the base of the blade. The tang and the point of the blade are missing. The sample was taken from the cutting edge.

HNY39b Museum Acc. D.390/1956-7 cp. -

A tanged, hooked cutting blade measuring 126mm long, and 31mm wide at the base of the blade. A rivet pierces the tang. The sample was taken from the cutting edge at the hook.

HNY40a Museum Acc. D.152/1957-8 (3) cp. -

A curved strip 149mm long, 5mm wide, and 3mm thick with an amorphous shaped bulb of iron at one end. The sample was taken from the flat end.

HNY40b Museum Acc. D.146/1957-8 cp. -

A straight bar measuring 478mm long, and 10x6mm in dimensions. One end of the bar has been forged an angle and pointed. The sample was taken from the pointed end.

HNY41a Museum Acc. D.314/1956-7 cp. -

A tanged blade measuring 158mm long, and 20mm wide at the base of the blade tapering to 11mm at the blunt point. The sample was taken from the cutting edge near the point.

HNY41b Museum Acc. D.302/1956-7 cp. -

A small, tanged blade measuring 98mm long, and 25mm wide at the base of the blade tapering to 15mm at the blunt point. The sample was taken from the cutting edge at the base of the blade.

HNY42a Museum Acc. D.152/1957-8 (4) cp. -

A curved rod measuring 178mm long, and 6x8mm in dimensions. One end of the rod was fashioned into a loop through which a bar still remains. The artefact was probably part of a bucket handle. The sample was taken from the fractured end of the rod.

HNY42b Museum Acc. D.145/1957-8 (1) cp. -

A chain link produced by forging a 426mm long, 4x5-5x10mm rod into an oval shape and then bending the longer sides of the oval together to form a figure 8. The artefact measures 185mm long, and 41-14-45mm wide. The sample was taken from one end of the artefact.

HNY43a Museum Acc. D.146/1957-8 (3) cp. -

A rod measuring 243mm long, and 7-9mm in diameter forged at an angle 6mm from one end. The sample was taken from the shaft near the bend.

HNY43b Museum Acc. 1940 Y cp. -

An adze measuring 157mm long, 9mm wide in front of the socket flaring to 36mm wide at the blade, and 28mm thick at the socket tapering to a sharp edge. The sample was taken from the blade.

HNY44a Museum Acc. D.155/1957-8 cp. -

A hooked strip fragment measuring 70mm long, 16-18mm wide, and 5-6mm thick. The sample was taken from the hooked end.

HNY44b Museum Acc. D.169/1957-8 cp. -

A large, tanged, hooked blade fragment measuring 109mm long, and 31mm wide at the base of the blade. The remains of a large rivet are present. The sample was taken from the inside edge near the fracture.

HNY45a Museum Acc. D.135/1957-8 cp. -

A ladle fragment measuring 283mm long, and 6mm in diameter along the shaft. The sample was taken from the spoon.

HNY45b Museum Acc. D.136/1956-7 cp. -

A straight rod measuring 291mm long and 9mm in diameter. A small loop 28mm in diameter is present at one end. The sample was taken from the shaft.

HNY46a Museum Acc. D.147/1957-8 cp. -

A curved rod measuring 296mm long, and 3x12mm in dimensions. A rivet hole pierces one end. The artefact was probably part of a bucket handle. The sample was taken from one edge.

HNY46b Museum Acc. D.369/1956-7 cp. -

A ring nave produced from a rod measuring 7-9mm in diameter. The diameter of the artefact is 98mm. The sample was taken from an external edge.

HNY47a Museum Acc. D.150/1957-8 (6) cp. -

A ring nave produced from a rod measuring 9-11mm in diameter. The external diameter of the artefact is 131mm. The sample was taken from an external edge.

HNY47b Museum Acc. D.150/1957-8 (4) cp. -

A ring nave produced from a rod measuring 8-12mm in diameter. The external diameter of the artefact is 134mm. The sample was taken from an external edge.

HNY48a Museum Acc. D.375/1956-7 cp. -

A cauldron hook 494mm long produced from a rod measuring 10mm in diameter. The sample was taken from the back of the loop.

HNY48b Museum Acc. D.170/1956-7 cp. -

A piece of forging scrap measuring 163mm long, 14mm wide at one end tapering to 6mm at the other, and 7mm thick. The sample was taken from the wider end.

HNY49a Museum Acc. D.170/1956-7 cp. -

A piece of forging scrap measuring 96mm long, 8-10mm wide, and 3-5mm thick. The sample was taken from the flattened end.

HNY49b Museum Acc. D.170/1956-7 cp. -

A piece of forging scrap measuring 138mm long, 6-12mm wide, and 2-7mm thick. The sample was taken from one edge of the artefact towards one end.

HNY50a Museum Acc. D.170/1956-7 cp. -

A piece of forging scrap measuring 197mm long, 7-14mm wide, and 4-6mm thick. The sample was taken from one edge of the artefact.

HNY50b Museum Acc. D.170/1956-7 cp. -

A piece of forging scrap measuing 142mm long, and 8-12mm in diameter. The sample was the corrosion part of one point.

HNY51a Museum Acc. D.170/1956-7 cp. -

A second sample was taken from the same point of the artefact described in HNY50b.

HNY51b Museum Acc. D.360/1956-7 cp. -

A ploughshare fragment measuring 134mm long, 36mm wide at the socket tapering to 22mm wide at the squared-off point, and 5mm thick. The sample was taken from the point.

HNY52a Museum Acc. D.140/1957-8 cp. -

A ploughshare measuring 163mm long, 41mm wide at the base of the socket tapering to a sharp point, and 8mm thick. The blade of the ploughshare was produced by incompletely welding two 4mm thick sheets of iron together. The sample was taken from the point.

HNY52b Museum Acc. D.358/1956-7 cp. -

Two ploughshares rusted together, one inserted into the socket of the other. The outer ploughshare measures 128mm long, and 44mm wide at the base of the socket tapering to a rounded point 23mm wide. The inner ploughshare measures 129mm long and 48mm wide at the base of the socket tapering to a point. The sample was taken from the point of the outer ploughshare.

HNY53a Museum Acc. D.113/1957-8 cp. -

A fragment measuring 96mm long, 39mm wide at the forged end tapering to 21mm at the roughly rounded point, and 24mm thick at the large end tapering to 6mm thick. The artefact could be the remains of a ploughshare. The sample was taken from the smaller end.

HNY53b Museum Acc. D.321/1956-7 cp. -

An adze measuring 158mm long, 50mm wide at the blade tapering to 19mm just in front of the socket, and 21mm thick at the socket tapering to a still sharp blade. The sample was taken from the blade.

HNY54a Museum Acc. D.323/1956-7 cp. -

An adze fragment measuring 120mm long, 49mm wide at the blade tapering to 13mm at the broken socket, and 25mm thick at the socket tapering to a still sharp blade. Most of the socket is missing. The sample was taken from the blade.

HNY54b Museum Acc. D.322/1956-7 cp. -

An adze measuring 129mm long, 29mm wide at the blade tapering to 12mm wide at the socket, and 13mm thick at the socket tapering to a still sharp blade. The sample was taken from the blade.

HNY55a Museum Acc. D.319/1856-7 cp. -

An adze measuring 185mm long, 51mm wide at the blade tapering to 18mm just above the socket, and 24mm thick at the socket tapering to a still sharp blade. The sample was taken from the blade.

HNY55b Museum Acc. D.335/1956-7 cp. -

A chisel measuring 179mm long, 14mm wide, and 1mm thick. The artefact shows indications of having been struck on the solid butt end. The sample was taken from the blade.

HNY56a Museum Acc. D.167/1957-8 (1) cp. -

An awl shaped rod measuring 84mm long, and 8mm in diameter at one end tapering to 5x2mm in dimensions at the point. The sample was taken from the point.

HNY56b Museum Acc. D.167/1957-8 (2) cp. -

A pointed rod measuring 85mm long, and 6-9mm in diameter. The sample was taken from the point.

HNY57a Museum Acc. D.385/1956-7 cp. -

A rod measuring 89mm long, and 12mm in diameter. One end has been intentionally ground to a pyramidal point within the last 30 years. The sample was taken from one side of the point.

HNY57b Museum Acc. D.126/1957-8 cp. -

A file-shaped bar measuring 126mm long, 15mm wide at one end tapering to 8mm at the other, and 4-8mm thick. The sample was taken from the wider end.

HNY58a Museum Acc. D.386/1956-7 cp. -

A rod measuring 78mm long, and 9x9mm in dimensions at one end tapering to a 4x2mm point at the other. The sample was taken from the point.

HNY58b Museum Acc. D.129/1957-8 cp. -

A tyre fragment measuring 168mm long, 32mm wide, and 2-4mm thick. The sample was taken from one side.

HNY59a Museum Acc. D.338/1956-7 cp. -

A socketed object measuring 157mm long, and 29mm in diameter at the base of the socket tapering to 10mm at the opposing end. The artefact may have been a chisel or a ferrule. The sample was taken from the end.

HNY59b Museum Acc. D.165/1957-8 (2) cp. -

A fragment measuring 77mm long, 6mm wide at one end tapering to 3mm at the centre then flaring to 13mm wide at the other end, and 2-4mm thick. The sample was taken from the wider end.

HNY60a Museum Acc. D.138/1957-8 cp. -

A head of a blacksmith's poker consisting of a 142mm long, 12mm diameter shaft attached to a flat, ovally shaped head 68mm long, 34mm wide, and 3mm thick. The sample was taken from the tip of the head.

HNY60b Museum Acc. D.158/1957-8 cp. -

A blacksmith's poker consisting of a 11x9mm shaft, 75mm long attached to a circular head 31mm in diameter, and 2mm thick. The sample was taken from the circular head.

HNY61a Museum Acc. D.391/1956-7 cp. -

A blacksmith's poker fragment consisting of an 11x5mm, 68mm long shaft attached to an ovally shaped head 115mm long, 42mm wide, and 3mm thick. The sample was taken from one edge of the head.

HNY61b Museum Acc. D.389/1956-7 cp. -

A blacksmith's poker fragment consisting of an 8mm diameter, 246mm long shaft attached to an ovally shaped head 154mm long, 43mm wide, and 3mm thick. The sample was taken from one edge of the head.

HNY62a Museum Acc. D.124/1957-8 cp. -

A socketed ploughshare measuring 599mm long, 47mm wide at the top of the socket, and 7mm thick. The sample was taken from one edge of the artefact towards the middle.

HNY62b Museum Acc. D.153/1957-8 cp. -

A roughly rectangular strip measuring 155mm long, 13-24mm wide, and 2mm thick. The sample was taken from one side near an end.

HNY63a Museum Acc. D.162/1957-8 cp. -

A poorly produced, socketed tool measuring 155mm long, and 28mm in diameter at the base of the socket tapering sharply to 12mm at the shaft. The artefact may have been a chisel. The sample was taken from the end of the shaft.

HNY63b Museum Acc. D.303/1956-7 cp. -

A shears blade measuring 112mm long, and consisting of a 2x14mm shaft attached to a 14mm wide blade tapering to a rounded point 9mm wide. The sample was taken from the cutting edge at the base of the blade.

HNY64a Museum Acc. D.309/1956-7 cp. -

A hooked fragment measuring 61mm long, 21mm wide at one end tapering to 6mm at the other, and 5-6mm thick. The sample was taken from the wider end.

HNY64b Museum Acc. D.336/1956-7 cp. -

A saw fragment measuring 121mm long, 31mm wide at the fracture tapering to 16mm at the rounded point, and 4mm thick. One tooth was removed for examination.

HNY65a Museum Acc. D.393/1956-7 cp. -

A blacksmith's poker fragment consisting of a 10x2mm, 157mm long shaft attached to an ovally shaped head 110mm long, 47mm wide, and 2mm thick. The sample was taken from one edge of the head.

HNY65b Museum Acc. D.133/1957-8 cp. -

A dagger measuring 219mm long, 35mm wide at the blade, and 2mm thick. The tip of the blade is missing. The sample was taken from one cutting edge towards the middle of the dagger.

HNY66a Museum Acc. D.3[illegible]/1956-7 cp. -

A tanged blade fragment measuring 78mm long, and 23mm wide at the base of the blade tapering to 18mm at the fracture. The point is missing. The sample was taken from the cutting edge near the fracture.

HNY66b Museum Acc. D.382/1956-7 cp. -

A clamp fragment produced from a rod measuring 147mm long, 3-8mm wide, and 4-9mm thick. The rod is bent in two places to form three sides of a rectangular outline. The sample was taken from one end.

HNY67a Museum Acc. D.366B/1956-7 cp. -

A tyre fragment measuring 628mm long, 34mm wide, and 5mm thick. The sample was taken from one side.

HNY67b Museum Acc. D.308/1956-7 cp. -

A hooked, blade fragment measuring 144mm long, and 13mm wide at the fracture tapering to 11mm at the rounded point. The sample was taken from the back edge at the hook.

HNY68a Museum Acc. D.326/1956-7 cp. -

A tanged file measuring 189mm long, and 8x10mm in dimensions at the top of the tang tapering to 4x3mm at the point. The teeth are evident on only one side, and set 1mm apart. The sample was taken from the tip.

HNY68b Museum Acc. D.143/1957-8 cp. -

A large, wedge shaped ingot measuring 126mm long, 41mm wide at the fracture tapering to a rounded point, and 68mm thick. The object could have been an anvil. The sample was taken from the point near one side.

HNY69a Museum Acc. D.144/1957-8 cp. -

An anvil measuring 178mm long, and 68x42mm in dimensions at the base tapering to a large, hooked, forged point. The sample was taken from one edge near the point.

HNY69b Museum Acc. D.189/1957-8 cp. -

A hooked rod measuring 86mm long, and 12mm in diameter. The sample was taken from the hooked end.

HNY70a Museum Acc. D.137/1957-8 cp. -

A hammer measuring 174mm long, 26mm wide at the socket tapering to 23mm at the head, and 10mm thick. The butt end of the artefact is missing. Two wedges are present in the socket. The sample was taken from one edge of the striking surface.

HNY70b Museum Acc. D.141/1957-8 cp. -

A brazing hammer measuring 143mm long, 7-10mm wide, and 11mm thick at the centre tapering to 9mm at the striking end and 3mm at the opposite end. The socket is 62mm from the striking head, and 68mm from the butt end. The sample was taken from just behind the striking head.

LSa Small Finds 773161 cp. -

A point measuring 60mm long, 25mm wide, and 12mm thick. The artefact may have been the tip of a ploughshare. The sample was taken from the point.

Unpublished.

LSb Small Finds 773162 cp. -

A leaf-shaped spearhead measuring 138mm long, and 21mm wide at the base of the blade tapering to a point. The artefact has a midrib and a long slender socket. Two grooves decorate the socket, and a rivet hole pierces the base. The sample was taken from the cutting edge.

Pit 365 Layer 366. Neal, 1979, Fig. 1.

LW1a cp. 5-7

A roughly rectangular fragment measuring 90mm long, 24-36mm wide, and 2mm thick. The sample was taken from one end.

Hut C8, wall trench. Drury, 1978a, Fig. 5.5.

LW1b cp. 5-7

A roughly rectangular fragment measuring 125mm long, 44-48mm wide and 1.5mm thick. The sample was taken from one end.

Hut C8, wall trench. Drury, 1978a, Fig. 5.7.

LW2a cp. 8

A strip fragment measuring roughly 44mm long, 30mm wide, and 2-3mm thick. The sample was taken from one end.

F189. Drury, 1978a, Fig. 5.28.

LW2b cp. 8

A roughly rectanglur plate measuring 37mm long, 12-30mm wide, and 9mm thick. The sample was taken from one end.

F232. Drury, 1978a, Fig. 5.30.

LW3a cp. 5-7

A fragment measuring 20-46mm long, 20-34mm wide, and 8mm thick. The artefact may be the tip of a ploughshare. The sample was taken from the point.

F255. Drury, 1978a, Fig. 5.22.

LW3b cp. 8

A sword scabbard fragment measuring 44-70mm long, 38mm wide, and 6mm thick. The sample was taken from one edge.

F277, posthole R7. Drury, 1978a, Fig. 5.29.

LW4a cp. 5-7

A rod measuring 58mm long, and 7x7mm in dimensions. The sample was taken from one end.

Hut C10, wall trench. Drury, 1978a, Fig. 5.11.

LW4b cp. 8

A large ring measuring 32mm in diameter, produced from an iron rod 9mm in diameter. A quarter of the ring is missing. The sample was taken from one of the fractures.

F171. Drury, 1978a, Fig. 5.26.

LW5a cp. 5-7

A bent nail shaft measuring 62mm long. The sample was taken from the top of the shaft.

Hut C14, wall trench. Drury, 1978a, Fig. 5.14.

LW5b cp. 5-7

A bucket handle attachment consisting of a plate measuring 79mm long, 22-30mm wide, and 2-10mm thick attached to an iron loop produced from a rod 6mm in diameter. Two rivet holes pierce the central axis. The sample was taken from one side of the plate near the bottom.

F253. Drury, 1978a, Fig. 5.21.

LW6a cp. 5-7

A clamp produced by bending a 116mm long, 5x10mm rod in two places to form a rectangular outline. The sample was taken from the fractured end.

Hut C9, wall trench. Drury, 1978a, Fig. 5.8.

LW6b cp. 5-7

A nail measuring 50mm long, with a small forged flattened head. The sample was taken from the point.

Hut C12, wall trench. Drury, 1978a, Fig. 5.13.

MH1a cp. -

An incomplete currency bar measuring 502mm long, 18mm wide at the base of the blade not tapering to the fracture, and 2mm thick. Wood is evident in the socket. The sample was taken from an edge near the base of the blade.

Unpublished.

MH1b cp. -

A currency bar blade measuring 377mm long, 17mm wide at the fracture tapering to 9mm at the point, and 2mm thick. The sample was taken from an edge near the base of the blade.

Unpublished.

MH2a cp. -

An incomplete currency bar measuring 343mm long, 19mm wide at the base of the blade, and 2mm thick. The sample was taken from an edge near the base of the blade.

Unpublished.

MH2b cp. -

A currency bar measuring 504mm long, 17mm wide at the base of the blade, and 2mm thick. Wood is evident in the socket. The sample was taken from an edge near the base of the blade.

Unpublished.

MH3a cp. -

A currency bar measuring 693mm long, 22mm wide at the base of the blade not tapering to the fracture, and 2mm thick. Wood is evident in the socket. The sample was taken from an edge near the base of the blade.

Unpublished.

MH3NDBY cp. -

A currency bar measuring 726mm long, 34mm wide at the base of the blade, and 5mm thick. The sample was taken from an edge near the base of the blade.

Unpublished.

ODF1a Small Finds 20 cp. SXN

A roughly rectangular strip measuring 35mm long, 10mm wide, and 3mm thick. The sample was taken from one end.

Pit 621. Unpublished.

ODF1b Small Finds 4 cp. 7

A small, socketed, hooked, cutting blade measuring 75mm long, and 18mm wide at the base of the blade. The sample was taken from the cutting edge at the hook.

Pit 387. Davies, 1980, Fig. 30-21.

ODF2a Small Finds 77 cp. -

A rod measuring 118mm long and 8mm in diameter. The artefact is roughly square at one end and flattened at the other. The sample was taken from the flattened end.

Pit 1185. Unpublished.

ODF2b Small Finds 14 cp. 1-3

An awl shaped rod measuring 8mm long, and 12mm in diameter at one end tapering to 3mm at the other. The sample was taken from the pointed end.

Pit 378. Davies, 1980, 114.

ODF3a Small Finds 45 cp. 8-9

A wire forged into a loop with both ends touching, but not welded. The wire is 5mm in diameter. The sample was taken from the base of the loop.

Pit 3470. Unpublished.

ODF3b Small Finds 75 cp. -

A rod fragment measuring 55mm long with one diamond shaped end tapering to a flat, curved strip. The sample was taken from the thick end.

Unpublished.

ODF4a Small Finds 74 cp. SXN

A tanged blade measuring 73mm long, and 15mm wide at the tang tapering to 7mm at the broken point. The end of the tang and the point are missing. The sample was taken from the cutting edge near the tang.

Main Ditch-NE corner. Unpublished.

ODF4b Small Finds 3 cp. 7

A point measuring 75mm long, 25mm wide at the fracture tapering to a point, and 10mm thick. The artefact may have been the tip of a ploughshare. The sample was taken from the rounded point.

Pit 387. Davies, 1980, Fig. 124.

ODF5a Small Finds 104 cp. 9

A slightly curved blade measuring 115mm long, and 22mm wide at the base of the blade tapering to a sharp point. One complete rivet and the remains of a second pierce the tang. The sample was taken from the inner curve of the blade.

Pit 2598. Davies, 1980, Fig. 30-22.

ODF5b Small Finds 70 cp. 7

A socketed chisel measuring 226mm long, and 18mm wide at the blade. The sample was taken from the blade.

Pit 2420. Davies, 1980, Fig. 29-14.

ODF6a Small Finds 67 cp. 7

A currency bar handle measuring 158mm long, 40mm wide at the base of the blade, and 4-5mm thick. The sample was taken from the remaining blade.

Pit 2420. Davies, 1980, 124.

ODF6b Small Finds 78 cp. -

A blade fragment measuring 45mm long, 25mm wide at the base of the blade tapering to 5mm at the point, and 3mm thick. The sample was taken from onc cnd.

Pit 2128. Davies, 1980, 114.

ODF7a Small Finds 2 cp. 8-9

A blade point measuring 95mm long, and 3mm wide at the fracture tapering to a still sharp point. The sample was taken from the cutting edge near the point.

Pit 547. Davies, 1980, 138.

ODF7b Small Finds 13 cp. -

A fragment measuring 60mm long, and 40mm wide at one end tapering to 15mm at the other. One rivet remains. The sample was taken from one edge near the base.

Pit 557 Layer 558. Unpublished.

ODF8a Small Finds 48 cp. SXN

A curved strip measuring 50mm long, 6mm wide, and 3mm thick. The sample was taken from one end.

Pit 621. Unpublished.

ODF8b Small Finds 1 cp. 8-9

A fragment measuring 30mm long, 22mm wide, and 3mm thick. The sample was taken from one edge.

Pit 547 Layer 560. Davies, 1980, 138.

ODF9a cp. -

A bronze covered circular object measuring 2mm thick, and 13mm in diameter.

Main enclosure ditch. Unpublished.

ODF9b Small Finds 136 cp. -

A nail measuring 36mm long with a forged flattened head. The sample was taken from the head.

Unpublished.

ODF10a cp. -

A fragment measuring 37mm long, 4-7mm wide, and 4-7mm thick. The sample was taken from one edge.

Pit 2676. Unpublished.

ODF10b Small Finds 73 cp. SXN

A socketed Saxon arrowhead measuring 107mm long, and 24mm wide at the blade tapering to a point. The blade possesses a midrib, and a rivet hole pierces the socket. The sample was taken from the blade.

Main Enclosure Ditch. Davies, 1979, Fig. 8-1.

ODF11a Small Finds 76 cp. 7

An awl shaped rod measuring 73mm long, and 4mm in diameter at one end tapering to a still sharp point at the other. The sample was taken from the blunt end.

Pit 2583. Unpublished.

ODF11b Small Finds 65 cp. 7

A currency bar blade measuring 150cm long, and 32mm wide at the fracture. The sample was taken from one edge.

Pit 2420 Layer 2577. Davies, 1980, 124.

ODF12a Small Finds 64 cp. 7

A large, socketed, hooked cutting blade measuring 150mm long and 38mm wide at the base of the blade. The point of the artefact is missing. One rivet pierces the socket between the two flanges. The sample was taken from the cutting edge at the hook.

Pit 2032. Davies, 1980, Fig. 30-18.

ODF12b Small Finds 54 cp. 7

A large socketed, tanged, hooked, cutting blade measuring 123mm long and 31mm wide at the base of the blade. The sample was taken from the cutting edge at the hook.

Pit 966. Davies, 1980, Fig. 38-37.

ODF13a Small Finds 108 cp. -

A roughly rectangular sheet measuring 52mm long, 20mm wide, and 3mm thick. The sample was taken from one end.

Pit 2789. Unpublished.

ODF13b Small Finds 74 cp. SXN

A roughly rectangular object measuring 32mm long, 18mm wide, and 2-3mm thick. The sample was taken from one end.

Main ditch-NE corner. Unpublished.

PTLDa cp. -

A 'spitsbarren' ingot measuring 603mm long, and 89x83mm in dimensions at the centre tapering to points on each side. The central portion measures 229mm long. The sample was taken from an edge of the central portion.

Tylecote, 1962, 210.

PTLDb cp. -

A 'spitsbarren' ingot measuring 521mm long, and 102x70mm in dimensions at the centre tapering to points on each side. The central portion measures 229mm long. The sample was taken from an edge of the central portion.

Tylecote, 1962, 210.

WD1a cp. 7

A currency bar measuring 585mm long, and 41mm wide at the base of the blade. The point of the artefact is missing. The sample was taken from an edge.

Hooley, 1921.

WD1b cp. 7

A currency bar blade measuring 597mm long, and 39mm wide at the base of the blade. The socket of the artefact is missing. The sample was taken from an edge.

Hooley, 1921.

WD2a cp. 7

A currency bar blade measuring 135mm long, and 38mm wide at the base of the blade. The socket and the point of the artefact are missing. The sample was taken from an edge.

Hooley, 1921.

WD2b cp. 7

A currency bar measuring 835mm long, and 40mm wide at the base of the blade. The artefact is bent near the tip. The sample was taken from an edge.

Hooley, 1921.

WD3a cp. 7

A currency bar measuring 820mm long, and 40mm wide at the base of the blade. The sample was taken from an edge.

Hooley, 1921.

WD3b cp. 7

A complete currency bar with a deformed socket, and bent and twisted towards the centre. The dimensions could not be measured accurately. The sample was taken from an edge.

Hooley, 1921.

WD4a cp. 7

A currency bar broken into two pieces and twisted and split towards the centre. The dimensions could not be measured accurately. The sample was taken from an edge.

Hooley, 1921.

WD4b cp. 7

A currency bar measuring 872mm long, and 41mm wide at the base of the blade. The sample was taken from an edge.

Hooley, 1921.

WD5a cp. 7

A currency bar blade measuring 697mm long, and 35mm wide at the base of the blade. The socket is missing. The sample was taken from an edge.

Hooley, 1921.

WD5b cp. 7

A currency bar blade measuring 346mm long, and 30mm wide at the fracture. The socket is missing. The sample was taken from an edge.

Hooley, 1921.

WD6a cp. 7

A currency bar blade measuring 461mm long, and 32mm wide at the fracture. The socket is missing. The sample was taken from an edge.

Hooley, 1921.

WD6b cp. 7

A currency bar blade measuring 240mm long, and 28mm wide at fracture. The point and socket are missing. The sample was taken from an edge.

Hooley, 1921.

WD7a cp. 7

A currency bar measuring 451mm long, and 32mm wide at the base of the blade. The artefact is bent and split along the length. The sample was taken from an edge.

Hooley, 1921.

WD7b cp. 7

A currency bar blade fragment measuring 157mm long, and 25mm wide at the fracture. The socket and point are missing. The sample was taken from an edge.

Hooley, 1921.

WD8a cp. 7

A currency bar blade measuring 226mm long, and 26mm wide at the fracture. The point and socket are missing. The sample was taken from an edge. The sample disintegrated during preparation.

Hooley, 1921.

WD8b cp. 7

A currency bar blade measuring 216mm long, and 27mm wide at the fracture. The socket and point are missing. The sample was taken from an edge.

Hooley, 1921.

WD9a cp. 7

A currency bar blade measuring 87mm long, and 35mm wide at fracture. The socket and point are missing. The sample was taken from an edge.

Hooley, 1921.

WD9b cp. 7

A currency bar broken into three pieces, measuring 100mm, 73mm, and 82mm long respectively. The sample was taken from an edge.

Hooley, 1921.

WD10a cp. 7

A currency bar blade measuring 376mm long, and 41mm wide at fracture. The socket and point are missing. The sample was taken from an edge.

Hooley, 1921.

WD10b cp. 7

A ploughshare point measuring 198mm long, 21mm wide at the fracture, and 5mm thick. The sample was taken from the point.

Hooley, 1921, 324.

WD11a cp. 7

A wedge measuring 94mm long, 20mm wide, and 14mm thick below the head tapering to a flat blade. The sample was taken from the blade.

Pit 2. Hooley, 1931, Fig. 45.

WD11b cp. 7

A rod measuring 98mm long, and 4mm in diameter at one end tapering to a flattened end. The sample was taken from the flattened end.

Pit 2, 3.5ft deep. Hooley, 1931, Fig. 46.

WD12a cp. 7

A blade fragment measuring 44mm long, and 20mm wide at the fracture tapering to a point. The tang is missing. The sample was taken from the cutting edge.

Pit 4, 2ft deep. Hooley, 1931, Fig. 83.

WD12b cp. 7

A linch-pin measuring 172mm long. The sample was taken from the shaft.

Pit 4. Hooley, 1931, Fig. 82.

WD13a cp. 7

A pointed bar measuring 120mm long, 30mm wide at the fracture, and 4-10mm thick. The sample was taken from the point.

Pit 4, Floor. Hooley, 1931, Fig. 47.

WD13b cp. 7

A bent rod measuring 102mm long and 2-4mm in diameter. The sample was taken from one end.

Hooley, 1931, Fig. 85.

WND1a Small Finds 630 cp. -

A rod measuring 175mm long, and 6-8mm in diameter with a hook at one end. The sample was taken from the base of the shaft.

Pit 9304 Layer 9331. Unpublished.

WND1b Small Finds 612 cp. -

A large socketed, hooked, cutting blade measuring 160mm long and 45mm wide at the base of the blade. A rivet pierces both flanges of the socket. The sample was taken from the cutting edge at the hook.

Pit 8083 Layer 8865. Unpublished.

WND2a Small Finds 635 cp. -

A blade measuring 174mm long, and 22mm wide at the base of the blade tapering to a point. The sample was taken from the cutting edge.

Pit 9127 Layer 9128. Unpublished.

WND2b Small Finds 617 cp. -

A nail measuring 37mm long with a small forged flattened head. The sample was taken from the head.

Pit 9951 Layer 9952. Unpublished.

WND3a Small Finds 618 cp. -

A nail measuring 49mm long with a large forged flattened head. The sample was taken from the head.

Pit 9914 Layer 9953. Unpublished.

WND3b Small Finds 631 cp. -

A nail measuring 34mm long with a small forged flattened head. The sample was taken from the head.

Pit 10093 Layer 10099. Unpublished.

WND4a Small Finds 647 cp. -

A small, hooked cutting blade measuring 100mm long, and 15mm wide at the base of the blade. A rivet protrudes from between the flanges of the socket. The sample was taken from the back of the blade at the hook

Pit 9105 Layer 9138. Unpublished.

WND4b Small Finds 661 cp. -

A nail measuring 39mm long with a small forged flattened head. The sample was taken from the head.

Pit 1989 Layer 2241. Unpublished.

WND5a Small Finds 662 cp. -

A nail measuring 49mm long with the head protruding to one side. The sample was taken from the head.

Pit 2066 Layer 2067. Unpublished.

WND5b Small Finds 663 cp. -

A nail measuring 45mm long with the head protruding to one side. The sample was taken from the head.

Pit 2199 Layer 1616. Unpublished.

WND6a Small Finds 668 cp. -

A nail measuring 54mm long with a forged flattened head. The sample was taken from the head.

Pit 9802. Unpublished.

WND6b Small Finds 526 cp. -

A nail shaft measuring 89mm long. The sample was taken from the end of the shaft.

Pit 6732 Layer 6723. Unpublished.

WND7a Small Finds 537 cp. -

A nail measuring 60mm long with a bent point, and a forged flattened head. The sample was taken from the head.

Pit 8817 Layer 8086. Unpublished.

WND7b Small Finds 540 cp. -

A nail measuring 75mm long with a forged flattened head. The sample was taken from the head.

Pit 8817 Layer 8080. Unpublished.

WND8a Small Finds 541 cp. -

A nail measuring 69mm long with a forged flattened head. The sample was taken from the head.

Pit 8817 Layer 8086. Unpublished.

WND8b Small Finds 542 cp. -

A rod measuring 105mm long, and 5x6mm in dimensions. The sample was taken from one end.

Pit 678 Layer 1787. Unpublished.

WND9a Small Finds 555 cp. -

A nail measuring 74mm long with a forged flattened head. The sample was taken from the head.

Pit 8817 Layer 8082. Unpublished.

WND9b Small Finds 561 cp. -

A nail measuring 117mm long with a forged flattened head. The sample was taken from the head.

Pit 8817 Layer 8086. Unpublished.

WND10a Small Finds 563 cp. -

A curved nail shaft measuring 54mm long with a forged flattened head. The sample was taken from the head.

Pit 8817 Layer 8086. Unpublished.

WND10b Small Finds 565 cp. -

A semi-cicular object measuring 2mm thick and 35mm in diameter. The sample was taken from the edge.

Pit 6732 Layer 1. Unpublished.

WND11a Small Finds 573 cp. -

A rougly rectangular strip measuring 38mm long, 16mm wide, and 5mm thick. The sample was taken from one end.

Pit 9806 Layer 9813. Unpublished.

WND11b Small Finds 576 cp. -

A rod measuring 86mm long, and 8mm in diameter at one end tapering to a 2x6mm flattened end. The sample was taken from the flattened end.

Pit 9806 Layer 9813. Unpublished.

WND12a Small Finds 544 cp. -

A nail measuring 50mm long with a forged flattened head. The sample was taken from the head.

Pit 8817 Layer 8086. Unpublished.

WND12b Small Finds 547 cp. -

A nail measuring 47mm long with a forged flattened head. The sample was taken from the head.

Pit 8817 Layer 8086. Unpublished.

WND13a Small Finds 549 cp. -

A nail measuring 48mm long with a forged flattened head. The sample was taken from the head.

Pit 8634 Layer 8635. Unpublished.

WND13b Small Finds 552 cp. -

A nail measuring 66mm long with a forged flattened head. The sample was taken from the head.

Pit 8817 Layer 8086. Unpublished.

WND14a Small Finds 604 cp. -

A bent nail measuring 51mm long with a thick square head. The sample was taken from the head.

Pit 201 Layer 208. Unpublished.

WND14b Small Finds 608 cp. -

A bent nail shaft measuring 51mm long. The sample was taken from the end of the shaft.

Pit 4719 Layer 2107. Unpublished.

APPENDIX C: THE RESULTS OF THE METALLURGICAL ANALYSES

This Appendix presents the data derived from the metallurgical analyses. The first category is equivalent to category 1 of Appendix A, and lists the Sample Number of each artefact. The second category is a brief description of the artefact sampled, with the third category listing the Ceramic Phase from which it dates. The following six categories present the percent of cobalt, nickel, copper, arsenic, phosphorus, and sulphur contained in each sample. It should be noted that concentrations are presented in categories four through nine which are below the Detection Limits presented in Figure 3.13. These numbers were employed to facilitate the computerized manipulation of the data only, and should not be considered as accurate readings. Category ten presents the density of slag and forging inclusions per sample in terms of "low" (under 2%), "medium" (2 to 4%), and "high" (over 4%). The eleventh column lists a relative guide of the grain size of the artefacts. The notation of "c.w." stands for cold-worked. Column twelve lists the hardness of the sample as determined by Vickers Hardness testing with a 5kg weight. The final category is the carbon concentration of each sample. Those carbon concentrations divided by the character "/" indicate samples which exhibited definite indivisible areas of differing carbon contents. The carbon concentrations divided by the character "-" indicate that the sample contained carbon contents varying within that range.

MOUNT	TYPE	C.P.	Co	Ni	Cu	As	P	S	%SLAG	GRAIN	VPN	%CARBON
ACC1a	CHISEL	1-3	-	-	-	-	-	-	-	-	-	NO METAL
ACC1b	SICKLE, SM.	1-3	0.003	0.046	-	-	0.011	0.004	LOW	LG	79.2	NONE
ACC2a	AWL/PUNCH	1-3	0.000	0.033	-	-	0.022	0.003	LOW	MED	260.0	0.5-0.6
ACC2b	AWL/PUNCH	1-3	0.001	0.042	-	-	0.011	0.011	LOW	SM-MED	179.0	0.2-0.3
ACC3a	BLADE	1-3	0.004	0.026	-	-	0.085	0.003	HIGH	SM	157.0	LOW
ACC3b	BLADE	1-3	0.066	0.084	-	-	0.031	0.005	-	-	-	NO METAL
ACC4a	BLADE	1-3	0.000	0.049	-	-	0.009	0.001	MED	MED	113.0	NONE
ACC4b	BLADE	1-3	0.001	0.012	-	-	0.008	0.038	-	-	-	NO METAL
ACC5a	BLADE	1-3	0.068	0.053	-	-	0.074	0.010	-	-	-	NO METAL
ACC5RD	SICKLE, LG.	-	0.008	0.022	-	-	0.303	0.007	-	-	-	NO METAL
B1a	SICKLE	9	0.009	0.014	-	-	0.195	0.007	HIGH	LG/SM	240.0	NONE/LOW
B1b	SICKLE	9	0.0[illegible]3	0.069	-	-	0.169	0.004	MED	LG/SM	113.0	NONE/LOW
B2a	SICKLE	9	0.005	0.021	-	-	0.014	0.001	LOW	SM-LG	133.0	0.2
B2b	SICKLE	9	0.005	0.027	-	-	0.025	0.001	LOW	LG	201.0	NONE
B3a	SICKLE	9	0.030	0.016	-	-	0.012	0.010	LOW	C.W.	116.0	0.2
B3b	HAMMER	9	0.006	0.041	-	-	0.101	0.008	LOW	SM	212.0	0.3
B4a	PICK-AXE, VERTICAL	9	0.004	0.089	-	-	0.213	0.011	MED	LG	122.0	NONE
B4b	PICK-AXE, HORIZONTAL	9	0.021	0.080	-	-	0.142	0.007	LOW	LG	137.0	NONE
B5a	AXE	9	0.000	0.029	-	-	0.114	0.004	MED	C.W.	177.0	LOW
B5b	PLOUGHSHARE	9	0.012	0.036	-	-	0.064	0.004	MED	LG/MED	138.0	NONE/0.2
B6a	PLOUGHSHARE	9	0.000	0.033	-	-	0.019	0.003	LOW	SM	142.0	0.3
B6b	CHISEL	9	0.041	0.055	-	-	0.204	0.006	LOW	LG	134.0	NONE
B7a	SICKLE	9	0.003	0.010	-	-	0.022	0.004	MED	MED	101.0	LOW
B7b	SICKLE	9	0.001	0.014	-	-	0.010	0.004	LOW	LG	177.0	NONE
B8a	SICKLE	9	0.001	0.018	-	-	0.116	0.005	LOW	LG	141.0	NONE
B8b	SICKLE	9	0.012	0.019	-	-	0.025	0.013	LOW	LG	125.0	0.2
B9a	SICKLE	9	0.039	0.152	-	-	0.101	0.004	LOW	MED	134.0	0.1
B9b	BLADE	9	0.025	0.042	-	-	0.096	0.010	LOW	SM	177.0	0.3
B10a	DAGGER	9	0.013	0.023	-	-	0.159	0.006	MED	LG	-	LOW
B10b	SPEARHEAD	9	0.001	0.017	-	-	0.143	0.007	LOW	LG	143.0	NONE
B11a	CAULDRON HOOK	9	0.000	0.024	-	-	0.007	0.007	MED	LG	132.0	0.4
B11b	BRIDLE BIT	9	0.000	0.036	-	-	0.027	0.001	LOW	LG	126.0	LOW-0.2
B12a	RING, BRONZED	9	0.003	0.014	-	-	0.019	0.004	MED	-	140.0	-
B12b	SLAVE CHAIN	9	0.012	0.049	-	-	0.121	0.005	LOW	-	146.0	-
B13a	HAMMER, BLADE	9	0.001	0.010	-	-	0.012	0.016	LOW	SM	317.0	0.5-0.6
B13b	HAMMER, BUTT	9	0.000	0.023	-	-	0.001	0.004	LOW	SM	295.0	LOW-0.6
BC1a	BLADE	4-7	0.001	0.030	-	-	0.024	0.012	LOW	MED	112.0	0.2/0.4
BC1b	SICKLE, LG.	4-7	0.004	0.046	-	-	0.463	0.004	HIGH	LG	240.0	NONE
BC2a	SICKLE	4-7	-	-	-	-	-	-	-	-	-	LOST
BC2b	SICKLE, LG.	4-7	0.000	0.019	-	-	0.165	0.001	MED	LG	143.0	NONE
BC3a	SICKLE, SM.	4-7	0.005	0.021	-	-	0.005	0.003	MED	MED	279.0	NONE
BC3b	SICKLE, SM.	4-7	0.000	0.010	-	-	0.003	0.006	HIGH	LG	118.0	NONE
BC4a	SPEARHEAD	4-7	0.016	0.026	-	-	0.101	0.037	MED	LG	288.0	NONE
BC4b	SPEARHEAD	4-7	0.004	0.034	-	-	0.147	0.006	LOW	LG/SM	137.0	NONE/0.2
BC5a	SPEARHEAD	4-7	0.006	0.034	-	-	0.100	0.014	MED	LG	186.0	NONE
BC5b	AWL/PUNCH	4-7	0.000	0.037	-	-	0.005	0.003	MED	SM	143.0	0.4
BC6a	ROD, FRAG.	4-7	0.133	0.141	-	-	2.109	0.016	-	-	-	NO METAL

MOUNT	TYPE	C.P.	Co	Ni	Cu	As	P	S	%SLAG	GRAIN	VPN	%CARBON
BC6b	ROD/AWL	4-7	0.001	0.009	-	-	0.030	0.011	MED	SM	138.0	0.2
BC7a	ANVIL, FRAG.	4-7	0.082	0.072	-	-	0.257	0.005	LOW	LG/SM	112.0	NONE/0.1
BC7WD	SICKLE	-	0.001	0.044	-	-	0.016	0.005	MED	C.W.	194.0	0.3
BH1a	ROD	9	0.003	0.019	-	-	0.111	0.008	MED	LG	165.0	NONE
BH1b	SCABBARD, FRAG.	6-7	0.006	0.035	-	-	0.233	0.019	HIGH	MED	144.0	NONE
BH2a	RING	9	0.001	0.023	-	-	0.065	0.008	MED	C.W.	126.0	NONE
BH2b	NAIL	9	0.001	0.012	-	-	0.045	0.063	HIGH	SM	138.0	NONE
BR1a	BLADE, FRAG.	7-8	0.015	0.051	-	-	0.026	0.015	HIGH	MED	190.0	NONE
BR1b	SHEET, FRAG.	7-8	0.005	0.011	-	-	0.038	0.060	-	-	-	NO METAL
BTC1a	SAW	7	0.001	0.014	-	-	0.062	0.020	MED	SM	133.0	NONE
BTC1b	SICKLE, SM.	7	0.010	0.166	-	-	0.012	0.007	MED	C.W.	124.0	0.2
BTC2a	BLADE	7	0.003	[illegible]	-	-	0.177	0.085	-	-	-	NO METAL
BTC2b	BLADE, FRAG.	7	0.004	0.048	-	-	0.164	0.004	LOW	LG/SM	136.0	NONE/LOW
BTC3a	BLADE, FRAG.	7	0.000	0.027	-	-	0.325	0.005	HIGH	LG	148.0	NONE
BTC3b	BLADE, FRAG.	7	0 021	0.060	-	-	0.020	0.010	HIGH	SM	131.0	NONE
BTH1a	BLADE	-	0.001	0.036	-	-	0.047	0.003	LOW	MED	241.0	NONE/0.5
BTH1b	SICKLE, SM.	-	0.001	0.030	-	-	0.102	0.009	MED	MED	144.0	0.1-0.2
C1a	SHEET, FRAG./SWORD	-	0.000	0.023	-	-	0.042	0.011	LOW	LG/SM	218.0	NONE/0.4
C1b	BLADE	-	0.000	0.035	-	-	0.025	0.005	LOW	LG	129.0	NONE
C2a	SHEARS	-	0.000	0.025	-	-	0.018	0.004	MED	LG	80.0	NONE
C2b	BLADE, FRAG.	-	0.014	0.039	-	-	0.049	0.008	LOW	SM	-	0.8
CC1a	AXE	8-9	-	-	-	-	-	-	-	-	-	LOST
CC1b	BLADE	8-9	0.040	0.022	-	-	0.028	0.012	-	-	-	NO METAL
CD1a	SICKLE, LG.	-	0.001	0.050	-	-	0.023	0.014	MED	C.W.	143.0	LOW
D1a	SPEARHEAD	7	0.000	0.019	0.005	0.025	0.013	0.001	2.9	MED	178.0	0.2
D1b	SPEARHEAD	6	0.004	0.049	0.003	0.004	0.018	0.001	2.0	MED	138.4	0.2
D2	SICKLE, LG.	7-8	0.000	0.013	0.000	0.018	0.225	0.008	5.0	MED	162.0	0.1
D3a	SPEARHEAD	-	0.001	0.029	0.000	0.005	0.360	0.010	3.4	LG	94.1	LOW
D3b	SICKLE, LG.	-	0.001	0.018	0.002	0.049	0.062	0.006	5.5	SM	186.6	0.1/0.7
D4a	BLADE	3	0.000	0.024	0.001	0.030	0.433	0.007	3.8	LG	159.5	NONE
D4b	BLADE	6	0.000	0.042	0.006	0.060	0.016	0.004	2.2	MED	98.8	NONE
D5a	SAW	-	0.003	0.047	0.004	0.007	0.113	0.003	2.1	MED	135.6	NONE
D5b	SICKLE, LG.	7	0.001	0.023	0.003	0.075	0.328	0.007	0.9	LG	142.5	NONE
D6	BLADE	SXN	0.009	0.041	-	-	0.071	0.008	2.8	SM	519.2	0.8
D7a	BLADE	7	0.000	0.022	-	-	0.017	0.004	0.5	C.W.	289.0	0.2/0.7
D7a	BLADE	7	0.224	0.183	-	-	0.013	0.004	0.5	C.W.	289.0	0.2/0.7
D7b	CHISEL	3	0.001	0.034	-	-	0.015	0.001	2.6	MED	-	NONE
D8a	SICKLE, LG.	-	0.000	0.025	-	-	0.775	0.005	-	-	-	NO METAL
D8b	SICKLE,SM.	7	0.003	0.020	-	-	0.064	0.008	2.4	LG	104.3	NONE
D9	SPEARHEAD/SAW	3	0.019	0.064	-	-	0.054	0.009	2.0	C.W.	173.8	0.1-0.2
D10	PLOUGHSHARE	7	0.009	0.032	-	-	0.525	0.008	4.5	LG	178.0	NONE
D11a	CHISEL	7	0.000	0.032	-	-	0.075	0.005	3.2	SM	161.5	0.2-0.4
D11b	CHISEL	7	0.004	0.043	-	-	0.157	0.006	1.3	MED	173.7	0.2/0.7
D12	BUCKET ATTACHMENT	6	0.000	0.022	-	-	0.564	0.022	1.5	LG	152.3	LOW
D13a	SICKLE, SM.	7	0.000	0.019	-	-	0.015	0.005	0.6	SM	217.3	0.2/0.8
D13b	SICKLE, LG., FRAG.	-	0.000	0.027	-	-	0.428	0.001	2.5	LG	216.0	LOW
D14a	BLADE	3	0.001	0.015	-	-	0.011	0.010	0.6	SM	161.0	0.5
D14b	SICKLE,SM.	8	0.001	0.021	-	-	0.333	0.014	1.2	LG	131.0	NONE

MOUNT	TYPE	C.P.	Co	Ni	Cu	As	P	S	%SLAG	GRAIN	VPN	%CARBON
D15a	SICKLE, LG.	7-8	0.012	0.003	-	-	0.091	0.006	1.4	SM	246.0	0.7
D15b	BLADE	7	0.000	0.032	-	-	0.040	0.004	3.8	MED	145.0	0.2
D16a	SICKLE, SM.	8	0.100	0.112	-	-	0.018	0.005	0.8	SM	160.0	0.3
D16b	SICKLE, SM.	6	0.000	0.025	-	-	0.023	0.009	2.1	MED	114.7	NONE
D17	SICKLE	3	0.420	0.382	-	-	0.060	0.004	0.9	LG	207.8	0.7
D18	BLADE	7	0.000	0.021	-	-	0.005	0.001	0.6	MED	154.5	NONE
D19	FINGER RING	8	0.003	0.064	-	-	0.649	0.003	5.3	LG	200.6	NONE
D20	STRIP	6	0.005	0.052	-	-	0.025	0.001	3.6	MED	-	NONE
D21a	TANG	7	0.026	0.052	-	-	0.065	0.006	0.7	LG/MED	96.8	NONE/LOW
D21b	STRIP	7	0.003	0.050	-	-	0.248	0.005	1.2	MED	127.0	NONE
D22a	STRIP	-	0.000	0.010	-	-	0.016	0.005	0.1	SM	130.4	0.6-0.8
D22b	STRIP	7	0.001	0.020	-	-	0.008	0.007	1.0	MED	130.0	0.6
D23	ROD	7	0.000	0.028	-	-	0.005	0.008	3.1	LG	-	NONE
D24a	OBJECT, SOCKETED	6/8	0.004	0.021	-	-	0.234	0.004	1.2	MED	109.8	LOW/0.6
D24b	STRIPS, RIVETED	7	0.000	0.025	-	-	0.033	0.003	0.2	SM	117.2	0.1
D25a	DISC, FLAT	7	0.001	0.014	-	-	0.034	0.003	0.5	MED	101.4	0.3/0.6
D25b	STRIP	3	0.042	0.075	-	-	0.043	0.006	1.2	MED	123.4	0.2-0.7
D26a	WEDGE	-	0.000	0.025	-	-	0.015	0.004	4.0	LG	70.9	NONE
D26b	BUCKET HANDLE	7	0.001	0.015	-	-	0.025	0.007	0.7	MED	108.2	0.2-0.4
D27a	SHEET	7	0.001	0.016	-	-	0.039	0.005	0.6	SM	112.0	0.2
D27b	CLAMP	7	0.004	0.022	-	-	0.033	0.008	0.6	LG/MED	94.6	LOW/0.3
D28a	TANG	7	0.010	0.014	-	-	0.016	0.001	0.4	SM	105.0	0.2
D28b	ROD	7	0.003	0.033	-	-	0.086	0.005	1.1	C.W.	145.6	LOW-0.2
D29a	CLAMP	6	0.000	0.019	-	-	0.028	0.006	0.2	LG	58.7	NONE
D29b	ROD	6	0.001	0.033	-	-	0.010	0.003	0.9	MED	120.8	0.2
D30a	STRIP	7	0.010	0.039	-	-	0.180	0.003	0.1	LG	103.3	NONE
D30b	INGOT	7	0.072	0.065	-	-	0.072	0.004	0.3	LG	78.5	NONE
D31a	FRAGMENT, TANGED	7	0.000	0.014	-	-	0.221	0.001	0.5	LG	105.9	NONE
D31b	SHEET	3?	-	-	-	-	-	-	-	-	-	NO METAL
D32a	ROD	3	0.030	0.058	-	-	0.010	0.003	0.3	LG	110.0	NONE
D32b	SHEET	1-3	0.000	0.015	-	-	0.005	0.004	0.7	SM	124.6	0.6/0.7
D33a	FRAGMENT, SOCKETED	5	0.001	0.013	-	-	0.106	0.001	2.6	MED	102.9	NONE-LOW
D33b	ROD	5	0.038	0.060	-	-	0.023	0.001	0.6	MED	73.6	NONE/0.2
D101	CURRENCY BAR	7	0.000	0.013	0.015	0.000	0.436	0.006	-	-	-	NONE
D102	CURRENCY BAR	7	0.034	0.086	0.000	0.013	0.824	0.005	-	-	-	NONE
D103	CURRENCY BAR	7	0.000	0.051	0.019	0.017	1.136	0.004	-	-	-	NONE
D104	CURRENCY BAR	7	0.000	0.031	0.025	0.000	0.279	0.008	-	-	-	NONE
D105	CURRENCY BAR	7	0.000	0.030	0.015	0.051	0.238	0.004	-	-	-	NONE
D106	CURRENCY BAR	7	0.000	0.021	0.047	0.016	0.224	0.014	-	-	-	NONE
D107	CURRENCY BAR	7	0.001	0.033	0.058	0.042	0.108	0.001	-	-	-	NONE
D108	CURRENCY BAR	7	0.000	0.029	0.050	0.038	0.331	0.004	-	-	-	NONE
D109	CURRENCY BAR	7	0.000	0.036	0.078	0.127	0.282	0.004	-	-	-	NONE
D110	CURRENCY BAR	7	0.001	0.034	0.018	0.060	1.069	0.001	-	-	-	NONE
D111	CURRENCY BAR	7	0.001	0.018	0.088	0.001	0.157	0.004	-	-	-	NONE
D112	CURRENCY BAR	7	0.000	0.027	0.087	0.028	0.203	0.004	-	-	-	NONE
D120	FRAGMENT, SOCKETED	7	0.000	0.011	0.089	0.000	0.049	0.006	-	-	-	-
D121	BLADE	7	0.000	0.026	0.034	0.020	0.017	0.032	-	-	-	-
D122	CURRENCY BAR, FRAG.	7	0.000	0.032	0.001	0.000	0.790	0.012	-	-	-	NONE

MOUNT	TYPE	C.P.	Co	Ni	Cu	As	P	S	%SLAG	GRAIN	VPN	%CARBON
D123	OBJECT, TANGED	7	0.779	0.103	0.000	0.025	0.031	0.01	-	-	-	-
D124	CLAMP	4	0.000	0.001	0.021	0.000	0.203	0.001	-	-	-	-
D125	ROD, FRAG.	7	0.000	0.052	0.019	0.119	0.269	0.000	-	-	-	-
D126	NAIL	7	0.000	0.032	0.024	0.207	0.025	0.006	-	-	-	-
D127	BAR	7	0.065	0.037	0.017	0.016	0.490	0.005	-	-	-	-
D128	RING STAPLE	7	0.000	0.001	0.024	0.000	0.344	0.071	-	-	-	-
D129	FERRULE	7	0.000	0.012	0.000	0.000	0.300	0.003	-	-	139.0	LOW-0.1
D130	SICKLE, SM.	7	0.000	0.028	0.000	0.001	0.030	0.003	-	-	122.0	0.1
D131	OBJECT	7	0.000	0.030	0.000	0.000	0.190	0.003	-	-	-	-
D132	ROD	7	0.562	0.112	0.010	0.000	0.067	0.016	-	-	-	-
D133	FILE	7	0.000	0.026	0.001	0.061	0.031	0.001	-	-	141.0	LOW
D134	FILE	7	0.000	0.156	0.000	0.042	0.007	0.001	-	-	120.0	LOW
D135	BLADE, FRAG.	7	0.000	0.021	0.000	0.000	0.006	0.003	-	-	-	-
D136	CURRENCY BAR, FRAG.	1-3	0.000	0.023	0.000	0.113	0.841	0.004	-	-	-	NONE
D137	FERRULE	7	0.062	0.035	0.000	0.352	0.221	0.006	-	-	-	-
D138	BUCKET ATTACHMENT	6	0.000	0.039	0.000	0.053	0.270	0.003	-	-	220.0	LOW-0.1
D139	CHISEL	5	0.000	0.020	0.000	0.019	0.018	0.006	-	-	329.0	0.1-0.8
D140	HORSESHOE	-	0.000	0.025	0.000	0.061	0.101	0.013	-	-	-	-
D141	TANG/WASHER	7	0.158	0.059	0.000	0.022	0.102	0.006	-	-	-	-
D142	BLADE	7	0.000	0.035	0.000	0.000	0.010	0.003	-	-	-	-
D143	RING	5	-	-	-	-	-	-	-	-	-	-
D144	BAR, THICK	7	0.024	0.051	0.000	0.000	0.308	0.004	-	-	-	-
D145	BAR	7	-	-	-	-	-	-	-	-	-	-
D146	BLADE	-	0.000	0.013	0.001	0.025	0.000	0.001	-	-	-	-
D147	BLADE	7	0.000	0.014	0.001	0.000	0.537	0.047	-	-	-	NO METAL
D148	NAIL	-	0.000	0.027	0.037	0.000	0.020	0.004	-	-	-	-
D149	NAIL	-	0.000	0.026	0.000	0.684	0.188	0.009	-	-	-	-
D150	NAIL	-	0.019	0.407	0.000	0.000	0.233	0.003	-	-	-	-
D151	WEDGE	7	0.000	0.042	0.029	0.079	0.011	0.001	-	-	-	-
D152	OBJECT	7	0.000	0.018	0.022	0.032	0.005	0.003	-	-	-	-
D153	OBJECT, TANGED	7	0.000	0.026	0.000	0.013	0.464	0.007	-	-	140.0	LOW
D154	SICKLE, LG.	3	0.000	0.134	0.035	0.013	0.442	0.005	-	-	230.0	LOW
D155	BAR	4-5	0.000	0.000	0.001	0.000	0.266	0.000	-	-	-	-
D156	BAR	4-5	0.000	0.021	0.000	0.053	1.085	0.003	-	-	-	-
D157	CHISEL	7	0.000	0.064	0.019	0.082	0.130	0.000	-	-	309.0	LOW-0.3
D158	CHISEL	7	0.000	0.017	0.000	0.021	0.740	0.004	-	-	415.0	0.3-0.4
D159	BAR, FRAG.	5	0.779	0.100	0.000	0.021	0.030	0.006	-	-	-	-
D160	STRIP	7	0.000	0.026	0.106	0.035	0.079	0.010	-	-	-	-
D161	BLADE, FRAG.	-	0.030	0.033	0.161	0.001	0.255	0.005	-	-	223.0	LOW
D162	BLADE, FRAG.	-	-	-	-	-	-	-	-	-	-	-
D163	PLOUGHSHARE	7	0.000	0.025	0.170	0.066	0.015	0.001	-	-	-	-
D164	ADZE	7	0.171	0.110	0.152	0.001	0.058	0.006	-	-	121.0	0.1-0.9
D165	SICKLE, LG.	7	-	-	-	-	-	-	-	-	-	NO METAL
D166	HAMMER	7	0.000	0.021	0.141	0.061	0.409	0.000	-	-	251.0	LOW
D167	HAMMER	7	0.000	0.025	0.131	0.046	0.146	0.000	-	-	-	-
D168	HAMMER, HEAVY	6	0.000	0.039	0.215	0.007	0.015	0.001	-	-	-	-
D169	HAMMER, HEAVY	6	1.893	0.231	0.000	0.032	0.019	0.001	-	-	235.0	0.5-0.8
D170	HAMMER, HEAVY	6	2.227	0.394	0.000	0.016	0.017	0.001	-	-	-	-

MOUNT	TYPE	C.P.	Co	Ni	Cu	As	P	S	%SLAG	GRAIN	VPN	%CARBON
D171	HAMMER, HEAVY	6	1.170	0.288	0.000	0.032	0.065	0.103	-	-	-	NO METAL
D172	FILE	7	0.000	0.018	0.000	0.031	0.021	0.006	-	-	225.0	0.3-0.4
D173	SAW	8	0.049	0.032	0.015	0.093	0.227	0.018	-	-	-	-
D174	SAW	8	0.000	0.061	0.000	0.212	0.035	0.004	-	-	174.0	LOW
D175	ROD	7	0.000	0.018	0.034	0.092	0.003	0.001	-	-	-	-
D176	CHISEL	7	0.000	0.028	0.000	0.000	0.033	0.012	-	-	-	-
D177	BAR	3	0.642	0.129	0.121	0.000	0.070	0.005	-	-	-	-
D178	RINGED ROD	-	0.031	0.046	0.025	0.001	0.542	0.007	-	-	-	-
D179a	PLATE	-	0.000	0.020	0.029	0.045	0.389	0.000	-	-	-	-
D179b	PLATE	-	0.000	0.026	0.000	0.000	0.498	0.000	-	-	-	-
D180	ROD	-	0.053	0.046	0.025	0.014	0.269	0.000	-	-	-	-
D181	ROD	-	0.001	0.024	0.018	0.025	0.449	0.013	-	-	-	-
D182	SICKLE	-	0.215	0.051	0.059	0.000	0.109	0.057	-	-	-	-
D183	SHEET	-	0.000	0.048	0.018	0.035	0.015	0.006	-	-	-	-
D184	CLAMP	-	3.310	0.818	0.026	0.016	0.083	0.000	-	-	-	-
D185	WEDGE	-	0.000	0.039	0.023	0.024	0.224	0.000	-	-	121.0	LOW
D186	WEDGE	-	0.000	0.050	0.001	0.021	0.342	0.016	-	-	272.0	0.3-0.5
D187	CAULDRON HOOK	-	0.021	0.035	0.176	0.024	0.085	0.005	-	-	-	-
D188	CAULDRON HOOK	-	0.041	0.043	0.147	0.049	0.126	0.005	-	-	-	-
D189	ROD	-	0.000	0.021	0.001	0.000	0.230	0.000	-	-	-	-
FBD1a	SICKLE, SM.	4-6	0.003	0.025	-	-	0.030	0.014	MED	MED	109.0	0.2
FBD1b	BLADE	4-6	0.038	0.085	-	-	0.344	0.018	MED	LG/MED	167.0	NONE/LOW
HNY1a	SPEARHEAD	-	0.000	0.019	-	-	0.073	0.036	-	-	-	NO METAL
HNY1b	SPEARHEAD	-	0.005	0.031	-	-	0.032	0.001	MED	SM/LG	199.8	NONE-0.4
HNY2a	SPEARHEAD	-	0.006	0.043	-	-	0.193	0.016	-	-	-	NO METAL
HNY2b	DAGGER	-	0.003	0.020	-	-	0.015	0.006	-	-	-	NO METAL
HNY3a	SPEARHEAD	-	0.001	0.014	-	-	0.152	0.013	-	-	-	NO METAL
HNY3b	SPEARHEAD	-	0.008	0.023	-	-	0.111	0.119	-	-	-	NO METAL
HNY4a	SPEARHEAD	-	0.000	0.039	-	-	0.033	0.003	HIGH	MED	159.0	NONE
HNY4b	ADZE	-	0.001	0.036	-	-	0.284	0.001	HIGH	LG	209.0	NONE
HNY5a	BLADE	-	0.003	0.026	-	-	0.417	0.019	-	-	-	NO METAL
HNY5b	CURRENCY BAR, FRAG.	-	0.003	0.024	-	-	0.147	0.006	MED	C.W.	172.0	NONE
HNY6a	CURRENCY BAR, FRAG.	-	0.004	0.065	-	-	0.976	0.003	HIGH	LG	232.4	NONE
HNY6b	BAR	-	0.004	0.038	-	-	0.181	0.006	HIGH	MED-LG	114.8	NONE
HNY7a	CURRENCY BAR, FRAG.	-	0.001	0.017	-	-	0.525	0.122	-	-	-	NO METAL
HNY7b	TYRE	-	0.000	0.023	-	-	0.010	0.006	HIGH	MED/LG	112.7	0.1/NONE
HNY8a	CURRENCY BAR, FRAG.	-	0.007	0.023	-	-	0.008	0.005	LOW	MED	91.5	NONE
HNY8b	CURRENCY BAR, FRAG.	-	0.007	0.046	-	-	0.562	0.001	HIGH	LG	230.2	NONE
HNY9a	CURRENCY BAR, FRAG.	-	0.001	0.018	-	-	0.467	0.001	HIGH	LG	179.0	NONE
HNY9b	CURRENCY BAR, FRAG.	-	0.006	0.038	-	-	0.688	0.010	MED	LG	229.0	NONE
HNY10a	PLOUGHSHARE	-	0.001	0.063	-	-	0.132	0.004	LOW	SM	168.7	NONE/0.1
HNY10b	CHISEL	-	0.044	0.050	-	-	0.109	0.008	MED	LG/MED	171.0	NONE/LOW
HNY11a	SPEARHEAD	-	0.007	0.046	-	-	0.038	0.006	MED	LG	105.5	NONE
HNY11b	SPEARHEAD	-	0.003	0.038	-	-	0.097	0.001	MED	MED	158.7	NONE/0.2
HNY12a	SPEARHEAD	-	0.001	0.013	-	-	0.045	0.019	-	-	-	NO METAL
HNY12b	PLOUGHSHARE	-	0.055	0.261	-	-	0.952	0.028	-	-	-	NO METAL
HNY13a	CURRENCY BAR, FRAG.	-	0.009	0.034	-	-	0.185	0.003	MED	LG	116.5	NONE
HNY13b	SPEARHEAD	-	0.003	0.014	-	-	0.646	0.029	-	-	-	NO METAL

MOUNT	TYPE	C.P.	Co	Ni	Cu	As	P	S	%SLAG	GRAIN	VPN	%CARBON
HNY14a	SPEARHEAD	-	0.000	0.022	-	-	0.011	0.001	LOW	MED	168.5	NONE
HMY14b	PLOUGHSHARE	-	0.055	0.021	-	-	0.225	0.004	MED	LG	144.3	NONE
HNY15a	PLOUGHSHARE	-	0.000	0.031	-	-	0.514	0.007	HIGH	LG	183.8	NONE
HNY15b	DAGGER	-	0.000	0.033	-	-	0.011	0.001	HIGH	SM	158.0	NONE
HNY16a	CURRENCY BAR	-	0.030	0.049	-	-	0.575	0.001	HIGH	LG	194.0	NONE
HNY16b	CURRENCY BAR, FRAG.	-	0.000	0.040	-	-	0.766	0.003	HIGH	LG	199.8	NONE
HNY17a	DAGGER	-	0.000	0.021	-	-	0.003	0.004	LOW	C.W.	215.5	0.2/0.3
HNY17b	SPEARHEAD	-	0.006	0.043	-	-	0.111	0.004	HIGH	MED	158.0	NONE
HNY18a	CHISEL	-	0.004	0.013	-	-	0.036	0.004	MED	SM	136.2	0.2
HNY18b	POKER	-	0.001	0.026	-	-	0.623	0.005	HIGH	LG	187.4	NONE
HNY19a	POKER	-	0.004	0.011	-	-	0.048	0.024	-	-	-	NO METAL
HNY19b	SPEARHEAD	-	0.005	0.062	-	-	0.034	0.004	LOW	C.W.	278.2	NONE
HNY20a	PICK/CHISEL	-	0.001	0.015	-	-	0.031	0.006	LOW	MED	185.6	0.3
HNY20b	SICKLE, SM.	-	0.001	0.011	-	-	0.170	0.051	-	-	-	NO METAL
HNY21a	CHISEL	-	0.000	0.032	-	-	0.021	0.003	MED	MED	117.0	0.1/0.5
HNY21b	CHISEL	-	0.001	0.023	-	-	0.018	0.001	HIGH	MED	108.0	0.1
HNY22a	SPEARHEAD	-	0.013	0.026	-	-	0.108	0.025	-	-	-	NO METAL
HNY22b	BLADE	-	0.001	0.023	-	-	0.171	0.035	-	-	-	NO METAL
HNY23a	CURRENCY BAR, FRAG.	-	0.000	0.033	-	-	0.282	0.007	MED	MED	140.4	NONE
HNY23b	CHISEL	-	0.000	0.027	-	-	0.013	0.008	MED	MED	215.4	0.6-0.7
HNY24a	ROD/AWL	-	0.000	0.032	-	-	0.023	0.001	HIGH	SM	119.4	LOW
HNY24b	AWL/PUNCH	-	0.025	0.017	-	-	0.020	0.007	LOW	SM	118.5	0.1-0.2
HNY25a	BLADE	-	0.000	0.033	-	-	0.071	0.010	LOW	SM/LG	113.0	0.1/NONE
HNY25b	TYRE	-	0.000	0.024	-	-	0.076	0.007	LOW	MED	124.0	0.3-0.4
HNY26a	SPEARHEAD	-	0.006	0.017	-	-	0.160	0.046	-	-	-	NO METAL
HNY26b	SPEARHEAD	-	0.006	0.028	-	-	0.266	0.020	-	-	-	NO METAL
HNY27a	PICK	-	0.001	0.017	-	-	0.397	0.004	MED	LG	190.6	NONE
HNY27b	SPEARHEAD	-	-	-	-	-	-	-	-	-	-	NO METAL
HNY28a	DAGGER	-	0.014	0.020	-	-	0.230	0.006	MED	MED	162.7	LOW
HNY28b	SAW	-	0.003	0.023	-	-	1.029	0.027	-	-	-	NO METAL
HNY29a	SICKLE, LG.	-	0.000	0.010	-	-	0.453	0.044	-	-	-	NO METAL
HNY29b	SICKLE, SM.	-	0.000	0.022	-	-	0.011	0.004	LOW	MED	282.3	0.7
HNY30a	DAGGER	-	0.005	0.017	-	-	0.318	0.016	-	-	-	NO METAL
HNY30b	SCABBARD, FRAG.	-	0.001	0.019	-	-	0.089	0.027	-	-	-	NO METAL
HNY31a	BLADE	-	0.014	0.049	-	-	0.227	0.008	MED	MED	167.3	NONE
HNY31b	LYNCH-PIN	-	0.004	0.043	-	-	0.024	0.003	LOW	SM	128.6	0.1
HNY32a	BLADE	-	0.007	0.042	-	-	0.101	0.012	LOW	SM	279.5	0.3-0.7
HNY32b	BLADE/SICKLE, LG.	-	0.008	0.055	-	-	0.103	0.008	MED	SM	-	0.2-0.3
HNY33a	FULLER	-	0.009	0.038	-	-	0.063	0.005	MED	MED	98.1	LOW
HNY33b	BLADE	-	0.003	0.025	-	-	0.018	0.013	MED	SM	152.7	0.2-0.3
HNY34a	BLADE	-	0.003	0.017	-	-	0.523	0.021	-	-	-	NO METAL
HNY34b	BLADE	-	0.001	0.010	-	-	0.305	0.040	-	-	-	NO METAL
HNY35a	STRIP	-	0.000	0.021	-	-	0.036	0.005	LOW	MED	192.5	NONE
HNY35b	SWORD	-	0.001	0.027	-	-	0.029	0.015	-	-	-	NO METAL
HNY36a	BLADE/SICKLE, SM.	-	0.000	0.029	-	-	0.072	0.005	LOW	LG	191.7	NONE
HNY36b	BLADE	-	0.006	0.031	-	-	0.173	0.024	HIGH	LG	130.0	NONE
HNY37a	BLADE	-	0.000	0.019	-	-	0.013	0.008	LOW	LG	77.3	NONE
HNY37b	BLADE	-	0.000	0.030	-	-	0.021	0.004	HIGH	MED	124.5	LOW-0.2

MOUNT	TYPE	C.P.	Co	Ni	Cu	As	P	S	%SLAG	GRAIN	VPN	%CARBON
HNY38a	BLADE	-	0.005	0.036	-	-	0.021	0.003	HIGH	MED	148.0	NONE
HNY38b	SICKLE, SM.	-	0.000	0.017	-	-	0.095	0.028	-	-	-	NO METAL
HNY39a	SICKLE, SM.	-	0.016	0.040	-	-	0.112	0.005	MED	SM	225.0	NONE/0.2
HNY39b	SICKLE, SM.	-	0.000	0.015	-	-	0.006	0.001	LOW	SM	131.3	0.4
HNY40a	OBJECT	-	0.000	0.027	-	-	0.012	0.005	LOW	SM	163.7	0.6
HNY40b	OBJECT/ROD	-	0.001	0.023	-	-	0.020	0.001	HIGH	MED	186.0	0.3
HNY41a	BLADE	-	0.001	0.016	-	-	0.116	0.007	-	-	-	NO METAL
HNY41b	BLADE	-	0.010	0.028	-	-	0.054	0.004	LOW	SM	141.0	LOW/0.2
HNY42a	OBJECT/BUCKET	-	0.007	0.030	-	-	0.108	0.004	HIGH	SM	257.8	LOW
HNY42b	LINK, LG.	-	0.000	0.018	-	-	0.008	0.001	LOW	MED	130.3	LOW
HNY43a	ROD	-	0.000	0.025	-	-	0.020	0.006	MED	SM	131.4	0.3/0.5
HNY43b	ADZE	-	0.000	0.017	-	-	0.029	0.005	LOW	MED	146.3	0.3-0.4
HNY44a	STRIP, FRAG.	-	0.029	0.102	-	-	0.129	0.000	LOW	MED	97.7	LOW
HNY44b	SICKLE, LG.	-	0.001	0.021	-	-	0.199	0.004	HIGH	LG	123.0	NONE
HNY45a	LADLE, FRAG.	-	0.001	0.024	-	-	0.039	0.004	-	-	-	NO METAL
HNY45b	ROD, LOOPED	-	0.006	0.024	-	-	0.091	0.066	-	-	-	NO METAL
HNY46a	OBJECT/BUCKET	-	0.000	0.021	-	-	0.359	0.100	-	-	-	NO METAL
HNY46b	RING-NAVE	-	0.001	0.022	-	-	0.014	0.001	LOW	LG	120.0	0.3-0.4
HNY47a	RING-NAVE	-	0.000	0.017	-	-	0.034	0.005	MED	SM	138.2	0.1
HNY47b	RING-NAVE	-	0.001	0.010	-	-	0.020	0.033	-	-	-	NO METAL
HNY48a	CAULDRON HOOK	-	0.001	0.024	-	-	0.079	0.004	HIGH	LG	118.3	0.1-0.2
HNY48b	SCRAP	-	0.001	0.024	-	-	0.022	0.003	MED	LG	206.2	0.3/0.5
HNY49a	SCRAP	-	0.001	0.023	-	-	0.036	0.005	LOW	SM	148.0	0.4
HNY49b	SCRAP	-	0.000	0.034	-	-	0.023	0.004	MED	SM	161.4	0.1-0.3
HNY50a	SCRAP	-	0.001	0.025	-	-	0.005	0.001	MED	MED	105.6	LOW-0.1
HNY50b	SCRAP	-	0.004	0.025	-	-	0.050	0.001	MED	MED	104.0	0.4
HNY51a	SCRAP	-	0.003	0.023	-	-	0.063	0.004	LOW	SM/LG	169.3	0.3/0.4
HNY51b	PLOUGHSHARE	-	0.010	0.029	-	-	0.117	0.007	HIGH	LG	113.0	NONE
HNY52a	PLOUGHSHARE	-	0.018	0.074	-	-	0.117	0.000	HIGH	LG	142.3	LOW-0.1
HNY52b	PLOUGHSHARE, DOUBLE	-	0.000	0.023	-	-	0.010	0.001	MED	LG	112.7	NONE
HNY53a	PLOUGHSHARE, FRAG.	-	0.004	0.037	-	-	6.723	0.007	LOW	MED	333.8	0.3-0.4
HNY53b	ADZE	-	0.016	0.024	-	-	0.139	0.005	HIGH	MED	128.0	NONE-LOW
HNY54a	ADZE	-	0.000	0.030	-	-	0.132	0.001	HIGH	SM/LG	108.1	NONE
HNY54b	ADZE	-	0.003	0.028	-	-	0.266	0.007	HIGH	SM	163.5	NONE-LOW
HNY55a	ADZE	-	0.007	0.081	-	-	0.028	0.005	LOW	MED	197.3	0.2-0.7
HNY55b	CHISEL	-	0.000	0.030	-	-	0.019	0.006	LOW	MED	146.5	0.2/0.5
HNY56a	ROD/AWL	-	0.001	0.043	-	-	0.014	0.005	MED	LG/SM	203.3	LOW/0.2
HNY56b	ROD	-	0.000	0.030	-	-	0.025	0.009	MED	SM	156.2	LOW
HNY57a	ROD	-	0.000	0.026	-	-	0.016	0.003	MED	MED	303.0	LOW
HNY57b	BAR	-	0.000	0.019	-	-	0.081	0.006	LOW	MED	209.5	0.1-0.3
HNY58a	ROD	-	0.001	0.021	-	-	0.025	0.005	LOW	SM	169.6	0.1-0.2
HNY58b	TYRE	-	0.003	0.012	-	-	0.202	0.029	-	-	-	NO METAL
HNY59a	CHISEL	-	0.001	0.019	-	-	0.005	0.003	LOW	LG	181.0	NONE
HNY59b	FRAGMENT	-	0.000	0.026	-	-	0.031	0.024	-	-	-	NO METAL
HNY60a	POKER	-	0.000	0.033	-	-	0.071	0.000	HIGH	LG/MED	138.6	NONE/0.1
HNY60b	POKER	-	0.000	0.023	-	-	0.045	0.003	LOW	LG	86.4	NONE
HNY61a	POKER	-	0.000	0.018	-	-	0.012	0.001	HIGH	SM	126.8	LOW-0.1
HNY61b	POKER	-	0.000	0.016	-	-	0.012	0.004	LOW	SM	181.0	0.6-0.7

MOUNT	TYPE	C.P.	Co	Ni	Cu	As	P	S	%SLAG	GRAIN	VPN	%CARBON
HNY62a	PLOUGHSHARE	-	0.003	0.028	-	-	0.094	0.003	MED	MED	123.0	LOW
HNY62b	STRIP	-	0.000	0.015	-	-	0.007	0.001	LOW	MED	203.3	0.6-0.7
HNY63a	CHISEL, FRAG.	-	0.001	0.026	-	-	0.147	0.007	MED	MED	134.3	LOW-0.1
HNY63b	SHEARS	-	0.009	0.022	-	-	0.044	0.009	-	-	-	NO METAL
HNY64a	FRAGMENT	-	0.004	0.032	-	-	0.040	0.047	-	-	-	NO METAL
HNY64b	SAW	-	0.007	0.022	-	-	0.010	0.003	HIGH	MED	117.0	LOW
HNY65a	POKER	-	0.006	0.035	-	-	0.090	0.001	HIGH	MED	113.5	LOW
HNY65b	DAGGER	-	0.001	0.029	-	-	0.012	0.006	LOW	SM	170.5	0.6
HNY66a	BLADE	-	0.056	0.078	-	-	0.077	0.003	MED	SM	172.5	0.3-0.5
HNY66b	CLAMP	-	0.000	0.011	-	-	0.013	0.003	LOW	MED	160.3	0.6
HNY67a	TYRE	-	0.000	0.019	-	-	0.068	0.001	HIGH	SM	181.3	0.5
HNY67b	SICKLE, FRAG.	-	0.000	0.025	-	-	0.025	0.007	MED	MED	125.8	NONE/0.2
HNY68a	FILE	-	0.004	0.067	-	-	0.071	0.007	MED	MED	200.8	0.5-0.6
HNY68b	INGOT/ANVIL	-	0.063	0.069	-	-	0.105	0.048	-	-	-	NO METAL
HNY69a	ANVIL	-	0.001	0.038	-	-	0.128	0.004	HIGH	MED	135.8	NONE-0.2
HNY69b	ROD	-	0.001	0.036	-	-	0.001	0.001	MED	SM	145.8	0.3
HNY70a	HAMMER	-	0.000	0.027	-	-	0.017	0.003	MED	SM	193.2	NONE-0.5
HNY70b	HAMMER	-	0.000	0.012	-	-	0.027	0.006	HIGH	LG	377.3	NONE
LSa	PLOUGHSHARE, FRAG.	-	0.022	0.025	-	-	0.905	0.004	MED	LG	238.4	NONE
LSb	SPEARHEAD	-	0.000	0.024	-	-	0.010	0.001	MED	SM	134.5	0.1/0.2
LW1a	STRIP	5-7	0.001	0.011	-	-	0.025	0.001	-	-	-	NO METAL
LW1b	PLATE/STRIP	5-7	0.001	0.023	-	-	0.122	0.037	-	-	-	NO METAL
LW2a	STRIP	8	0.001	0.040	-	-	0.175	0.005	LOW	MED	129.0	LOW-0.1
LW2b	PLATE	8	0.019	0.046	-	-	0.780	0.019	HIGH	LG	168.0	NONE
LW3a	BAR	5-7	0.026	0.064	-	-	0.173	0.026	-	-	-	NO METAL
LW3b	SWORD SCABBARD	8	0.001	0.034	-	-	0.033	0.025	-	-	-	NO METAL
LW4a	ROD	5-7	0.003	0.010	-	-	0.065	0.004	LOW	SM	179.0	0.1
LW4b	RING	8	0.004	0.027	-	-	0.148	0.009	MED	MED	182.0	NONE
LW5a	NAIL SHANK	5-7	0.043	1.001	-	-	0.055	0.067	-	-	-	NO METAL
LW5b	BUCKET HANDLE MOUNT	5-7	0.008	0.022	-	-	0.035	0.029	-	-	-	NO METAL
LW6a	CLAMP	5-7	0.026	0.062	-	-	0.152	0.014	-	-	-	NO METAL
LW6b	NAIL SHANK	5-7	0.004	0.208	-	-	0.032	0.060	MED	C.W.	226.0	NONE
MH1a	CURRENCY BAR	-	0.000	0.021	-	-	0.019	0.008	MED	MED	301.0	NONE
MH1b	CURRENCY BAR	-	0.000	0.016	-	-	0.017	0.009	MED	C.W.	248.5	LOW
MH2a	CURRENCY BAR	-	0.000	0.021	-	-	0.020	0.010	MED	SM	86.4	NONE
MH2b	CURRENCY BAR	-	0.000	0.022	-	-	0.024	0.007	MED	C.W.	108.7	NONE
MH3a	CURRENCY BAR	-	0.000	0.026	-	-	0.056	0.010	LOW	SM	169.3	0.1-0.2
MH3NDB	CURRENCY BAR	-	0.005	0.034	-	-	0.088	0.008	HIGH	C.W.	28.3	NONE/0.2
ODF1a	STRIP	-	0.003	0.020	-	-	0.038	0.007	LOW	MED	146.4	0.4/0.2
ODF1b	SICKLE, SM.	-	0.000	0.045	-	-	0.283	0.004	MED	MED	-	NONE
ODF2a	ROD	-	0.028	0.052	-	-	0.021	0.003	HIGH	C.W.	170.8	LOW
ODF2b	ROD/AWL	-	0.004	0.035	-	-	0.232	0.006	HIGH	MED	158.7	NONE
ODF3a	WIRE	-	0.004	0.024	-	-	0.014	0.010	MED	MED	184.6	NONE-0.1
ODF3b	ROD/OBJECT	-	0.001	0.019	-	-	0.128	0.004	HIGH	SM	133.0	NONE/0.3
ODF4a	BLADE, FRAG.	SXN	0.005	0.019	-	-	0.094	0.071	LOW	SM	301.0	0.7
ODF4b	PLOUGHSHARE	-	0.008	0.030	-	-	0.752	0.008	HIGH	LG	239.4	NONE
ODF5a	BLADE	-	0.051	0.022	-	-	0.003	0.003	MED	LG	175.3	0.3/0.6
ODF5b	CHISEL	-	0.000	0.018	-	-	0.017	0.023	-	-	-	NO METAL

MOUNT	TYPE	C.P.	Co	Ni	Cu	As	P	S	%SLAG	GRAIN	VPN	%CARBON
ODF6a	CURRENCY BAR, FRAG.	-	0.000	0.021	-	-	0.870	0.006	HIGH	LG	226.0	NONE
ODF6b	BLADE, FRAG.	-	0.000	0.010	-	-	0.040	0.025	-	-	-	NO METAL
ODF7a	BLADE, FRAG.	-	0.008	0.042	-	-	0.024	0.013	MED	SM	197.6	0.3
ODF7b	FRAGMENT	-	0.008	0.023	-	-	0.035	0.014	MED	LG/SM	104.2	NONE/0.4
ODF8a	STRIP	-	0.001	0.029	-	-	0.068	0.001	HIGH	MED	218.6	NONE
ODF8b	FRAGMENT	-	0.000	0.012	-	-	0.046	0.043	-	-	-	NO METAL
ODF9a	CIRCULAR OBJ., BRONZED	-	0.008	0.044	-	-	0.034	0.005	MED	-	-	-
ODF9b	NAIL	-	0.003	0.032	-	-	0.051	0.004	MED	C.W.	150.0	NONE
ODF10a	ROD	-	0.179	0.111	-	-	0.029	0.006	LOW	LG	165.2	0.5
ODF10b	ARROWHEAD	SXN	0.001	0.034	-	-	0.103	0.013	LOW	MED	124.7	NONE/0.2
ODF11a	AWL/PUNCH	-	0.015	0.041	-	-	0.167	0.001	HIGH	LG	263.3	NONE
ODF11b	CURRENCY BAR	-	0.000	0.030	-	-	0.343	0.005	MED	MED	160.7	NONE
ODF12a	SICKLE, LG.	-	0.014	0.143	-	-	0.702	0.001	HIGH	LG	222.6	NONE
ODF12b	SICKLE, LG.	-	0.003	0.034	-	-	0.417	0.005	MED	LG/SM	199.8	NONE/LOW
ODF13a	SHEET	-	0.004	0.073	-	-	0.137	0.011	HIGH	MED	200.6	LOW
ODF13b	STRIP, FRAG.	-	0.000	0.031	-	-	0.261	0.009	HIGH	MED	250.5	NONE
PTLDa	INGOT	-	0.000	0.019	-	-	0.009	0.007	MED	MED	106.0	0.1
PTLDb	INGOT	-	0.006	0.026	-	-	0.010	0.009	MED	MED	86.2	NONE
WD1a	CURRENCY BAR	7	0.000	0.027	-	-	0.191	0.001	LOW	MED	189.0	NONE
WD1b	CURRENCY BAR	7	0.001	0.046	-	-	0.478	0.003	HIGH	LG	195.0	NONE
WD2a	CURRENCY BAR	7	0.003	0.036	-	-	0.458	0.005	MED	LG	255.0	NONE
WD2b	CURRENCY BAR	7	0.001	0.023	-	-	0.120	0.006	LOW	LG	224.0	NONE
WD3a	CURRENCY BAR	7	0.001	0.025	-	-	0.214	0.005	HIGH	LG	160.0	NONE
WD3b	CURRENCY BAR	7	0.013	0.039	-	-	0.132	0.001	MED	MED	179.0	NONE
WD4a	CURRENCY BAR	7	0.000	0.024	-	-	0.535	0.009	MED	LG	217.0	NONE
WD4b	CURRENCY BAR	7	0.001	0.016	-	-	0.104	0.005	MED	C.W.	316.0	NONE
WD5a	CURRENCY BAR	7	0.010	0.041	-	-	0.438	0.001	LOW	LG	302.0	NONE
WD5b	CURRENCY BAR	7	0.006	0.032	-	-	0.346	0.001	MED	SM	178.0	LOW
WD6a	CURRENCY BAR	7	0.012	0.082	-	-	0.601	0.003	HIGH	LG	209.0	NONE
WD6b	CURRENCY BAR	7	0.012	0.064	-	-	0.543	0.003	HIGH	LG	193.0	NONE
WD7a	CURRENCY BAR	7	0.001	0.020	-	-	0.129	0.008	MED	MED	135.0	NONE
WD7b	CURRENCY BAR	7	0.006	0.050	-	-	0.381	0.003	LOW	MED	217.0	NONE
WD8a	CURRENCY BAR	7	-	-	-	-	-	-	-	-	-	LOST
WD8b	CURRENCY BAR	7	0.018	0.079	-	-	0.855	0.003	HIGH	LG	228.0	NONE
WD9a	CURRENCY BAR	7	0.015	0.094	-	-	0.774	0.001	HIGH	MED	171.0	NONE
WD9b	CURRENCY BAR	7	0.003	0.053	-	-	0.674	0.006	HIGH	LG	183.0	NONE
WD10a	CURRENCY BAR	7	0.000	0.019	-	-	0.194	0.004	HIGH	C.W.	166.0	NONE
WD10b	PLOUGHSHARE	7	0.000	0.020	-	-	0.903	0.001	HIGH	LG	220.0	NONE
WD11a	WEDGE	7	0.001	0.017	-	-	0.259	0.006	MED	MED	333.0	0.3/0.7
WD11b	ROD	7	0.023	0.102	-	-	0.202	0.006	MED	LG/MED	203.0	NONE/0.2
WD12a	BLADE, FRAG.	8-9	0.036	0.023	-	-	0.043	0.017	-	-	-	NONE
WD12b	LYNCH-PIN	8-9	0.031	0.086	-	-	0.035	0.005	MED	SM	107.0	NONE/0.1
WD13a	BAR	7	0.003	0.014	-	-	0.785	0.017	HIGH	LG	252.0	NONE
WD13b	ROD	8-9	0.085	0.060	-	-	0.132	0-013	MED	C.W.	197.0	LOW
WND1a	ROD	-	0.006	0.053	-	-	0.524	0.003	LOW	LG	225.8	NONE
WND1b	SICKLE, LG.	-	0.001	0.043	-	-	0.005	0.003	LOW	C.W.	124.0	LOW-0.2
WND2a	BLADE	-	0.000	0.022	-	-	0.181	0.027	-	-	-	NO METAL
WND2b	NAIL	-	0.014	0.068	-	-	0.174	0.001	HIGH	SM	118.5	NONE

MOUNT	TYPE	C.P.	Co	Ni	Cu	As	P	S	%SLAG	GRAIN	VPN	%CARBON
WND3a	NAIL	-	0.051	0.117	-	-	0.447	0.006	LOW	SM	192.0	NONE
WND3b	NAIL	-	0.046	0.068	-	-	0.028	0.007	LOW	LG	79.2	NONE
WND4a	SICKLE, SM.	-	0.001	0.018	-	-	0.869	0.044	-	-	-	NO METAL
WND4b	NAIL	-	0.004	0.029	-	-	0.339	0.017	MED	MED	131.0	NONE
WND5a	NAIL	-	0.009	0.030	-	-	0.290	0.014	LOW	C.W.	-	NONE
WND5b	NAIL	-	0.000	0.008	-	-	0.016	0.003	LOW	LG	110.6	NONE
WND6a	NAIL	-	0.000	0.010	-	-	0.073	0.006	HIGH	SM	235.4	0.2-0.3
WND6b	NAIL	-	0.014	0.042	-	-	0.026	0.008	MED	MED	176.0	NONE
WND7a	NAIL	-	0.020	0.047	-	-	0.224	0.007	LOW	MED	164.7	NONE
WND7b	NAIL	-	0.001	0.016	-	-	0.045	0.008	LOW	SM	103.4	NONE
WND8a	NAIL	-	0.000	0.017	-	-	0.012	0.011	MED	MED	81.4	NONE
WND8b	ROD/NAIL	-	0.005	0.009	-	-	0.046	0.001	HIGH	MED	157.5	NONE
WND9a	NAIL	-	0.000	0.005	-	-	0.311	0.008	-	-	-	NO METAL
WND9b	NAIL, LG.	-	0.109	0.094	-	-	0.040	0.003	HIGH	LG	80.4	NONE
WND10a	NAIL	-	0.011	0.049	-	-	0.074	0.001	MED	MED	112.0	NONE-LOW
WND10b	OBJECT	-	0.009	0.033	-	-	0.081	0.001	HIGH	SM	207.5	NONE
WND11a	STRIP	-	0.001	0.024	-	-	0.005	0.001	MED	MED	142.0	NONE
WND11b	ROD/NAIL	-	0.000	0.015	-	-	0.009	0.001	LOW	SM	142.0	NONE
WND12a	NAIL	-	0.001	0.027	-	-	0.016	0.001	LOW	MED	77.3	NONE
WND12b	NAIL	-	0.008	0.027	-	-	0.067	0.001	MED	LG	95.7	NONE
WND13a	NAIL	-	0.001	0.034	-	-	0.053	0.007	LOW	C.W.	126.0	NONE
WND13b	NAIL	-	0.004	0.016	-	-	0.150	0.004	MED	MED	102.0	NONE
WND14a	NAIL	-	0.000	0.025	-	-	0.084	0.004	MED	LG	197.0	NONE
WND14b	NAIL	-	0.001	0.030	-	-	0.024	0.001	HIGH	MED	141.5	NONE

BIBLIOGRAPHY

AGRICOLA, G.
1950: (trans. Hoover, C. & Hoover, L.H.) *De Re Metallica* (New York).

ALLEN, D.
1967: 'Iron Currency Bars in Britain'. *Proceedings of the Prehistoric Society*, 33, 307-335.

AMERICAN SOCIETY FOR TESTING AND MATERIALS.
1983: *1983 Annual Book of ASTM Standards* (Philadelphia).

ANNABLE, F.K.
1958: 'Excavation and Field-work in Wiltshire, 1957'. *Wiltshire Archaeological Magazine*, 57, 206, 2-17.
1961: 'Excavation and Fieldwork in Wiltshire, 1960'. *Wiltshire Archaeological Magazine*, 58, 209, 30-8.

BALCH, H.E.
1914: *Wookey Hole, its Caves and Cave Dwellers* (Oxford).

BERSU, G.
1940: 'Excavations at Little Woodbury, Wiltshire, Part 1'. *Proceedings of the Prehistoric Society*, 6, 30-111.

BIEK, L.
1982: 'The Ethnic Factor in Archaeotechnology'. In Aspinall, A. and Warren, S.E. (editors), *Proceedings of the 22nd Symposium on Archaeometry, University of Bradford* (Bradford), 303-315.

BIRINGUCCIO, V.
1942: (trans. Smith, C.S.) *The Pirotechnia* (New York).

BOYLSTON, H.M.
1928: *An Introduction to the Metallurgy of Iron and Steel* (New York).

BRADLEY, R.
1984: *The Social Foundations of Prehistoric Britain* (London).

BRAILSFORD, J.W.
1948: 'Excavations at Little Woodbury, Wiltshire (1938-39), Part 2'. *Proceedings of the Prehistoric Society*, 14, 1-23.
1949: 'Excavations at Little Woodbury, Parts 4 & 5'. *Proceedings of the Prehistoric Society*, 15, 156-168.

BRANDES, E.D. (editor)
1983: Smithells Metals Reference Book (Sixth edition, London).

BREWER, C.W.
1976: 'Metallographic examination of six ancient steel weapons'. Journal of the Historical Metallurgy Society, 10, 1, 1-10.

BROUGH, B.H.
1906: 'The Early Use of Iron'. Journal of the Iron and Steel Institute, 69, 233-253.

BULLEID, A. and GRAY, H.
1911: The Glastonbury Lake Village Vol. 1 (Taunton).
1917: The Glastonbury Lake Village Vol. 2 (Taunton).
1948: Meare Lake Village Vol. 1 (Taunton).
1953: Meare Lake Village Vol. 2 (Taunton).

BURGESS, C.
1979: 'A find from Boyton, Suffolk, and the end of the Bronze Age in Britain and Ireland'. In Burgess, C. and Coombs, D. (editors), Bronze Age Hoards: Some Finds Old and New (Oxford, BAR British Series 67).

BURROW, I.
1982: 'Hillforts and Hilltops 1000 BC-1000 AD'. In Aston, M. & Burrow, I. (editors), The Archaeology of Somerset (Taunton).

BUSHE-FOX, J.P.
1915: Excavations at Hengistbury Head, Hampshire in 1911-12 (Oxford, Society of Antiquaries Research Report 3).

CALEY, E.R.
1964: Analysis of Ancient Metals (Oxford).

CAMP, J.M. and FRANCIS, C.B.
1920: The Making, Shaping and Treating of Steel (Second edition, Pittsburgh).

CHAMPION, T.C.
1975: 'Britain in the European Iron Age'. Archaeologia Atlantica, 1, 2, 127-145.
1980: 'The early development of iron-working'. Nature, 284, 513-514.

CHAMPION, T.C. and CHAMPION, S.T.
1981: 'The Iron Age in Hampshire'. In Shennan, S.J. and Schadla-Hall, R.T. (editors), The Archaeology of Hampshire (Winchester).

CHARLES, J.A.
1973: 'Heterogeneity in Metals'. Archaeometry, 15, 1, 105-114.

CHILDE, V.G.
1942: What Happened in History (London).
1949: Prehistoric Communities of the British Isles (Third edition, London).

CLARK, G.
1948: Prehistoric England (Fourth edition, London).

CLAY, R.C.C.
1924: 'An Early Iron Age Site on Fifield Bavant Down'. Wiltshire Archaeological Magazine, 42, 140, 457-496.

CLEERE, H.F.
1972: 'The Classification of Early Iron-Smelting Furnaces'. Antiquaries Journal, 52, 8-23.
1976: 'Some operating parameters for Roman ironworks'. Bulletin of the Institute of Archaeology, London, 13, 233-246.
1981: The Iron Industry of Roman Britain (London Ph.D. thesis).

COGHLAN, H.H.
1941: 'Prehistoric Iron prior to the Dispersion of the Hittite Empire'. Man, 59, 74-80.
1956: Notes on Prehistoric and Early Iron in the Old World (Oxford, Pitt Rivers Museum Occasional Papers on Technology No 8).
1956-7: 'A Note upon Iron as a Material for the Celtic Sword'. Sibrium, 3, 129-136.

COLES, J.
1982: 'Prehistory in the Somerset Levels 4000-100 BC'. In Aston, M. and Burrow, I. (editors), The Archaeology of Somerset (Taunton).

COLLINS, J.H.
1969: A Handbook to the Minerology of Cornwall and Devon (Truro).

COLLIS, J.
1984: The European Iron Age (London).

CRAWFORD, O.G.S. & WHEELER, R.E.M.
1921: 'The Llynfawr and other Hoards of the Bronze Age'. Archaeologia, 71, 133-140.

CREW, P.
1984: Bryn y Castell 1984 Interim Report (Maentwrog, Gwynedd).

CUNLIFFE, B.W.
1966: Regional Groupings within the Iron Age of Southern Britain (Cambridge Ph.D. thesis).
1976: 'The Origins of Urbanization in Britain'. In Cunliffe, B.W. and Rowley, T. (editors), Oppida in Barbarian Europe (Oxford, BAR Supplementary Series 11).
1978a: Iron Age Communities in Britain (Second edition, London).
1978b: 'Settlement and Population in the British Iron Age: Some Facts, Figures and Fantasies'. In Cunliffe, B.W. and Rowley, T. (editors), Lowland Iron Age Communities in Europe (Oxford, BAR Supplementary Series 48).
1981a: 'A century of Hampshire archaeology'. In Shennan, S.J. and Schalda-Hall, R.T. (editors), The Archaeology of Hampshire (Winchester).
1981b: 'Money and Society in pre-Roman Britain'. In Cunliffe, B.W. (editor), Coinage and Society in Britain and Gaul: some current problems (London, CBA Research Report 38).
1982a: 'Britain, The Veneti and Beyond'. Oxford Journal of Archaeology, 1, 1, 39-68.
1982b: 'Iron Age Settlements and Pottery 650 BC-AD 60'. In Aston, M. and Burrow, I. (editors), The Archaeology of Somerset (Taunton).
1984a: Danebury: An Iron Age Hillfort in Hampshire Vol. 1 and 2 (London, CBA Research Report 52).
1984b: 'Iron Age Wessex: continuity and change'. In Cunliffe, B. and Miles, D. (editors), Aspects of the Iron Age in Central Southern Britain (Oxford, University of Oxford Committee for Archaeology Monograph No 2).

CUNLIFFE, B.W and ORTON, C.
1984: 'The Radiocarbon Dating'. In Cunliffe, B.W., Danebury: An Iron Age Hillfort in Hampshire Vol. 1 (London, CBA Research Report 52).

CUNNINGTON, M.E.
1923a: The Early Iron Age Inhabited Site at All Cannings Cross (Devizes).
1923b: 'Pits in Battlesbury Camp'. Wiltshire Archaeological Magazine, 42, 139, 368-373.
1930: 'Romano-British Wiltshire'. Wiltshire Archaeological Magazine, 45, 153, 166-216.

CUNNINGTON, B.H. and CUNNINGTON, M.E.
1913: 'Casterly Camp Excavations'. Wiltshire Archaeological Magazine, 38, 119, 53-105.

DAVIES, D.J. and OELMANN, L.A.
1983: The Structure, Properties and Heat Treatment of Metals (London).

DAVIES, S.M.
1979: 'Excavations of Old Down Farm, Andover, Part I: Saxon'. Proceedings of the Hampshire Field Club, 36, 161-180.
1980: 'Excavations at Old Down Farm, Andover, Part II: Prehistoric and Roman'. Proceedings of the Hampshire Field Club, 37, 81-163.

DRURY, P.
1978a: Excavations at Little Waltham, 1970-71 (London, CBA Research Report 26).
1978b: 'Little Waltham and Pre-Belgic Iron Age Settlement in Essex'. In Cunliffe, B.W. and Rowley, T. (editors), Lowland Iron Age Communities in Europe (Oxford, BAR Supplementary Series 48).

DRYDEN, H.E.L.
1885: 'Hunsbury or Danes Camp, and the Discoveries there'. Associated Architectural Societies Reports and Papers, 18, 53-61.

EHRENREICH, R.M. and SALTER, C.
1983: 'The Metallurgy and Chemistry of Some Iron Artefacts from the Iron Age Hillfort at Danebury, Hampshire'. Presented at the 23rd International Symposium on Archaeometry.

FASHAM, P.J.
1978: 'An Iron Age site on Winnal Down'. In Fasham, P.J. (editor), M3 Archaeology 1977 (Winchester).

FELL, C.I.
1937: 'The Hunsbury Hillfort, Northants'. Archaeological Journal, 93, 57-100.
1953: 'Hunsbury Hill'. Archaeological Journal, 110, 212-213.

FORBES, R.J.
1950: Metallurgy in Antiquity (Leiden).

FOX, C.
1945: A Find of the Early Iron Age from Llyn Cerrig Bach, Anglesey: An Interim Report (Cardiff).
1946: A Find of the Early Iron Age from Llyn Cerrig Bach, Anglesey (Cardiff).

FOX, C. and HYDE, H.
1939: 'A second Cauldron and an Iron Sword from the Llyn Fawr Hoard, Rhigos, Glamorganshire.' Antiquaries Journal, 19, 4, 369-404.

FRIEND, J.N.
1926: Iron in Antiquity (London).

GEORGE, T.J.
1917: 'Hunsbury'. Journal of the Northamptonshire Natural History Society, 18-19, 1-42.

GINGELL, C.
1979: 'The Bronze and Iron Hoard from Melksham and another Wiltshire Find'. In Burgess, C. and Coombs, D. (editors), Bronze Age Hoards - Some Finds Old and New (Oxford, BAR British Series 67).

GOODMAN, W.L.
1962: The History of Woodworking Tools (London).

GRAY, H. and COTTON, M.
1966: The Meare Lake Village Vol. 3 (Taunton).

GREENWOOD, W.H.
1884: Steel and Iron (New York).

GROVES, A.W.
1952: Wartime Investigations into the Haematite and Manganese Ore Resources of Great Britain and Northern Ireland (London, Ministry of Supply, Permanent Records of Research and Development Monograph 20-703).

HALDANE, J.W.
1970: 'A study of the chemical composition of pre-Roman ironwork from Somerset'. Bulletin of the Historical Metallurgy Group, 4, 2, 53-66.

HARBORD, F.W.
1904: The Metallurgy of Steel (London).

HARDING, D.W.
1972: The Iron Age in the Upper Thames Basin (Oxford).
1974: The Iron Age in Lowland Britain (London).
1979: Celts in Conflict: Hillfort Studies 1927-77 (Edinburgh).

HAWKES, C.F.C.
1936: 'The Excavations at Buckland Rings, Lymington, 1935'. Proceedings of the Hampshire Field Club, 13, 124-164.
1940: 'The Excavations at Bury Hill, 1939'. Proceedings of the Hampshire Field Club, 14, 291-337.

HAWKES, J. and HAWKES, C.F.C.
1953: Prehistoric Britain (Hamondsworth).

HAWKES, C.F.C. and SMITH, M.A.
1957: 'On some Buckets and Cauldrons of the Bronze and Early Iron Ages'. Antiquaries Journal, 37, 3, 131-198.

HAWLEY, W.
1924: 'Romano-British Villages on Upavon and Rushall Downs'. Wiltshire Archaeological Magazine, 42, 138, 277-230.

HEDGES, R.E.M. and SALTER, C.J.
1979: 'Source Determination of Iron Currency Bars through Analysis of the Slag Inclusions'. Archaeometry, 21, 2, 161-175.

HIGGINS, R.A.
1983: Engineering Metallurgy 1. Applied Physical Metallurgy (Fifth edition, London).
1984: Engineering Metallurgy 2. Metallurgical Process Technology (Fifth edition, London).

HODGES, T.R.
1906: 'Meon Hill and its Treasures'. Transactions of the Birmingham Archaeological Society, 32, 101-115.

HOGG, A.H.A.
1979: British Hillforts: An Index (Oxford, BAR British Series 62).

HONEYCOMBE, R.W.K.
1981: Steels: Microstructure and Properties (London).

HOOLEY, R.W.
1921: 'Note on a Hoard of Iron Currency Bars found on Worthy Down, Winchester'. Antiquaries Journal, 1, 321-327.
1931: 'Excavation of an Early Iron Age Village on Worthy Down, Winchester'. Proceedings of the Hampshire Field Club, 10, 178-192.

HOYT, S.L.
1954: ASME Handbook: Metals Properties (New York).

JESSUP, R.F. and COOK, N.C.
1936: 'Excavations at Bigberry Camp, Harbledown'. Archaeologia Cantiana, 48, 151-168.

KARSTEN, K.J.B.
1841: Handbuch der Eisenhuttenkunde (Berlin).

KNIGHT, D.
1984: Late Bronze Age and Iron Age Settlement in the Nene and Great Ouse Basins (Oxford, BAR British Series 130).

KNOX, R.
1963: 'Detection of Iron Carbide Structure in the Oxide Remains of Ancient Steel'. Archaeometry, 6, 43-45.

LANG, J. and WILLIAMS, A.R.
1975: 'The Hardening of Iron Swords'. Journal of Archaeological Science, 2, 3, 199-207.

MACGREGOR, M. and SIMPSON, D.D.A.
1963: 'A Group of Iron Objects from Barbury Castle, Wilts'. Wiltshire Archaeological Magazine, 58, 211, 394-402.

MANNING, W.H. and SAUNDERS, C.
1972: 'A Socketed Iron Axe from Maids Moreton, Buckinghamshire, with a note on the type'. Antiquaries Journal, 52, 276-292.

MARTIN, G. and FOLEY, V.
1979: 'Weyland the Smith: some findings'. Journal of the Historical Metallurgy Society, 13, 1, 38-39.

MARYON, H.
1938: 'Some Prehistoric Metalworkers' Tools'. Antiquaries Journal, 18, 243-250.

MCGRATH, J.N.
1968: 'A preliminary report on the metallographic examination of four fragmentary Early Iron Age sword blades from Llyn Cerrig Back, Anglesey'. Bulletin of the Historical Metallurgy Group, 2, 2, 78-80.

MERCER, H.C.
1929: Ancient Carpenters' Tools (Doylestown, Penn.).

MYERS, J.
1922: 'Note on Pre-Roman Iron Bars'. Journal of Chemical Industry, 41, 9, 133T-134T.

NEAL, D.S.
1979: 'Bronze Age, Iron Age, and Roman settlement sites at Little Somborne and Ashley, Hampshire'. Proceedings of the Hampshire Field Club, 36, 91-144.

NORTHOVER, P.
1982: 'The Metallurgy of the Wilburton Hoards'. Oxford Journal of Archaeology, 1, 1, 69-109.
1984: 'Iron Age bronze metallurgy in Central Southern England'. In Cunliffe, B. and Miles, D. (editors), Aspects of the Iron Age in Central Southern Britain (Oxford, University of Oxford Committee for Archaeology Monograph No 2).

PERCY, J.
1864: Metallurgy: the art of extracting metals from their ores, and adapting them to various purposes of manufacture (London).

PIASKOWSKI, J.
1974: 'Metallographic Examinations of Ancient Iron Objects'. In Gosh, A.K. (editor), Perspectives in Paleoanthropology (Calcutta).

PIGGOTT, S.
1949: British Prehistory (London).

PUGH, R.B. and CRITTALL, E. (editors)
1957: A History of Wiltshire Vol. 1, Part 1 (Oxford).

REGIONI, A. and GARINO, G.
1955: 'Esame Technologico di un Gruppo di Spade Galliche della Lombardia Nord-Occidentale'. Sibrium, 2, 43-55.

RICHARDSON, M.T.
1889a: Practical Blacksmithing Vol. 1 (New York).
1889b: Practical Blacksmithing Vol. 2 (New York).
1890: Practical Blacksmithing Vol. 3 (New York).
1891: Practical Blacksmithing Vol. 4 (New York).

ROLLASON, E.C.
1973: Metallurgy for Engineers (Fourth edition, London).

ROWLANDS, M.J.
1980: 'Kinship, alliance, and exchange in the European Bronze Age.' In Barrett, J. and Bradley, R. (editors), Settlement and Society in the British Later Bronze Age (Oxford, BAR British Series 83(i)).

RUDDOCK, F.A.
1961: 'Metal-working at Bagendon and Finds of Metal, Clay, Stone, Flint and Glass'. In Clifford, E.M., Bagendon: A Belgic Oppidum (Cambridge).

SALTER, C.
1984: 'Interim Report on the Ironwork from Danebury'. In Cunliffe, B.W., Danebury: An Iron Age Hillfort in Hampshire Vol. 2 (London, CBA Research Report 52).

SALTER, C. and EHRENREICH, R.M.
1984: 'Iron Age iron metallurgy in Central Southern Britain'. In Cunliffe, B. & Miles, D. (editors), Aspects of the Iron Age in Central Southern Britain (Oxford, University of Oxford Committee for Archaeology Monograph No 2).

SAUVER, A.
1935: The Metallography and Heat Treatment of Iron and Steel (Fourth edition, New York).

SCOTT, B.G.
1974: 'Notes on the development of metallographic studies of ancient iron'. Journal of the Historical Metallurgy Society, 8, 2, 88-91.

SELLWOOD, L.
1984: 'The Iron Artefacts'. In Cunliffe, B.W., Danebury: An Iron Age Hillfort in Hampshire Vol. 2 (London, CBA Research Report 52).

SMITH, R.A.
1905: 'The Ancient British Iron Currency'. Proceedings of Society of Antiquaries, 20, 179-194.

SMITHELLS, C.J.
1967: Metals Reference Book Vol. 3 (Fourth edition, New York).

SMYTHE, J.A.
1936-7: 'Ancient Sussex Iron Blooms: A Metallographic Examination'. Transactions of the Newcomen Society, 37, 197-203.

SPRATLING, M.G., et al.
1980: 'An Iron Age Bronze Foundry at Gussage All Saints, Dorset: Preliminary Assessment of Technology'. In Slater, E.A. and Tate, J.O. (editors), Proceedings of the 16th International Symposium on Archaeometry and Archaeological Prospection, Edinburgh 1976 (Edinburgh).

STEAD, J.E.
1918 'Note on inclusions in steel and ferrite lines'. Journal of the Iron and Steel Institute, 97, 1, 287-292.

STRAKER, E.
1931: Wealden Iron (London).

TACITUS, P.C.
1948: (trans. Mattingly, H.) Germania (London).

THEOPHILUS, R.
1847: (trans. Murray, H.) Upon Various Arts (London).

THOMPSON, F.H.
1984: 'Excavations at Bigberry, Near Canterbury, 1978-80'. Antiquaries Journal, 63, 2, 237-278.

TYLECOTE, R.F.
1962: Metallurgy in Archaeology (London).

WAINWRIGHT, G.J.
1977: 'A Celtic Farmstead in Southern Britain'. Scientific American, 237, 156-169.

WAINWRIGHT, G.J. (continued)
1979: *Gussage All-Saints: An Iron Age Settlement in Dorset* (London).

WELLS, P.S.
1980: *Culture Contact and Culture Changes: Early Iron Age central Europe and the Mediterranean World* (New York).
1984: *Farms, Villages, and Cities* (Ithaca).

WHEELER, R.E.M. and WHEELER, T.V.
1932: *Report on the Eacavation of Prehistoric, Roman and post-Roman Site in Lydney Park, Gloucestershire* (Oxford).

WHITELEY, J.H.
1918: 'The Effect of Cold-work on the Divorce of Pearlite'. *Journal of the Iron and Steel Institute*, 1, 353-360.
1920: 'The Distribution of Phosphorus in Steel between the points AC1 and AC3'. *Journal of the Iron and Steel Institute*, 51, 1, 359-382.
1921: 'Cupric etching effects produced by phosphorus and oxygen in iron'. *Journal of the Iron and Steel Institute*, 53, 1, 277-289.

WYNNE, E.J. and TYLECOTE, R.F.
1958: 'An experimental investigation into primitive iron smelting techniques'. *Journal of the Iron and Steel Institute*, 190, 339-348.